SOLAR II

How to Design, Build and Set Up Photovoltaic Components and Solar Electric Systems

by PHILLIP HURLEY

Wheelock Mountain Publications
is an imprint of
Good Idea Creative Services
Wheelock VT
USA

www.goodideacreative.com

First print edition 2012

Notice of Rights

Wheelock Mountain Publications is an imprint of:

Good Idea Creative Services
324 Minister Hill Road
Wheelock VT 05851 USA

ISBN 978-0-9837847-3-9

Library of Congress Control Number: 2012941688

Library of Congress subject headings:

Photovoltaic power generation --Amateurs' manuals.
Photovoltaic power generation --Handbooks, manuals, etc.
Photovoltaic power systems --Amateurs' manuals.
Photovoltaic power systems --Handbooks, manuals, etc.
Solar energy --Amateurs' manuals.
Solar energy--Handbooks, manuals, etc.

Disclaimer

The reader of this book assumes complete personal responsibility for the use or misuse of the information contained in this book. The information in this book may not conform to the reader's local safety standards. It is the reader's responsibility to adjust this material to conform to all applicable safety standards after conferring with knowledgeable experts in regard to the application of any of the material given in this book. The publisher and author assume no liability for the use of the material in this book as it is for informational purposes only. If these terms are not acceptable to you, then don't read this book.

Contents

Contents

Contents

Contents

Contents

Batteries (continued)

Preface

The Solar II project demonstrates how to build, set up and configure a variety of photovoltaic system components. It is a presentation of basic concepts of design for small solar electric power systems, and an introduction to the tools and components needed.

This text begins where *Build Your Own Solar Panel* leaves off. You have your photovoltaic panels, whether purchased or made yourself, and now you need to complete your PV system, so that you can plug in and use that solar power.

Basics of Electricity for PV Systems

Characteristics and measurement of electricity

In order to understand the components of a solar electric system, a basic knowledge of electricity is needed. There are two types of electricity that the PV power system installer will be working with: alternating current (AC) and direct current (DC).

Direct current flows in one direction only. This is the type of electricity that PV cells produce. Batteries in PV systems store DC electricity and can be charged by solar panels.

Alternating current alternates its direction of flow. It is the type of electricity provided by local power companies in most cases. This electricity is usually stepped down on a pole transformer near your house or place of business to provide 115 to 120 volts or 240 volts, and alternates at 60 cycles per second in the USA. In Europe, the AC alternates at 50 cycles per second. It is generated by large dynamos at hydroelectric, fossil-fuel fired, or atomic energy power plants.

Most appliances used in homes and workplaces use AC electricity. Inverters are used in photovoltaic systems to convert the DC electricity from the battery to AC so that common AC appliances can be run from the system. DC appliances can be run directly from batteries.

Measuring electricity – voltage and current

Both DC and AC electricity have common characteristics: voltage and current. For a basic understanding of these characteristics, consider a wire conductor to be similar to a pipe carrying water.

The voltage is the water pressure and the current is similar to the amount of water that passes out of the pipe in a specific period of time. Current is measured in amperes, and voltage in volts.

Since both AC and DC electricity must pass through some form of conductor to do useful work, another important consideration is resistance. Resistance is a reduction in current flow, and every conductor has some measure of resistance to it. Going back to the water pipe analogy, consider two pipes, one large and one small. The large pipe causes less resistance and the smaller pipe causes more resistance. Resistance is measured in ohms.

Watts and joules

The rate of the electricity – how fast it is being used or supplied – is measured in watts, which indicates the power of a system. One watt is an energy flow of one joule per second. A joule is equal to one newton of force exerted over a distance of one meter. Kinetically, this is the amount of force it takes to accelerate a mass of one kilogram so that it picks up one meter per second of velocity during each second that the force is exerted. Work and energy use the same unit, of measurement, the joule.

To further illustrate, consider that voltage is a measure of how many electrons are flowing through a circuit. This flow is measured with a unit called the coulomb. One coulomb equals 1.6×10^9 electrons. A voltage of one volt indicates that one joule of energy will be available for each coulomb that flows through a circuit.

Current is measured in amps which is how many coulombs per second are flowing past any given point in a circuit. A current reading of one amp indicates that 1.6×10^{19} electrons per second are flowing past any given point in a circuit.

VOLTS	x	AMPS	=	WATTS
(joules/coulomb)		(coulombs/second)		(joules/second)

Rather than using the joule, power companies (who are more interested in calculating rate of use) measure power generation and consumption in terms of kilowatt hours (kWh). A kilowatt hour equals 3600 kj. For instance, an appliance rated at 1000 watts used for one hour will use one kilowatt hour of energy, or 3600 kj.

Since one watt equals one joule per second, and we are using 1000 watts per second for 3600 seconds, we are using 3,600,000 joules per hour, or one kilowatt hour of electricity. Since the number of joules used is so enormous, it is more convenient for power companies to bill by the kilowatt hour rather than state your energy usage for the month in joules.

To illustrate the relationship between joules and kilowatt hours, consider the use of a 75 watt light bulb for one hour. This bulb will use 75 Wh (watt hours) or 0.075 kWh of energy. To convert this into joules, multiply 75 x 3600, which equals 270 kj.

Using Ohm's law

Most calculations that you will need to perform will be related to the relationship between volts, amperes and resistance in your PV generating and power distribution system, and the appliances used. To perform these calculations, you apply what is known as Ohm's law.

The particular formulas related to Ohm's law use the symbol I to represent current, the symbol R to represent resistance, the symbol E to represent voltage, and the symbol P to represent Power. Power (P) is usually measured in units of watts. Current (I) is measured in units of amperes, resistance (R) is measured in units of ohms, and voltage (E) is measured in units of volts.

To calculate the resistance of a circuit or wire run, and you know the voltage and current: **R = E/I**

To calculate the current in a circuit, and you know the voltage and resistance: **I = E/R**

To calculate the voltage in a circuit, and you know the current and resistance: **E = IR**

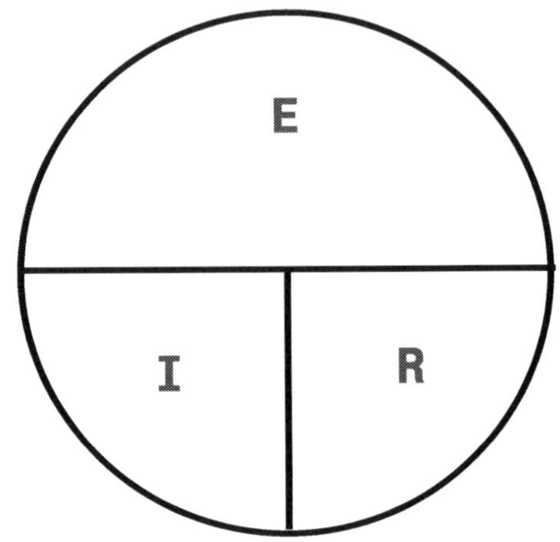

To easily remember which operation to perform, use the simple ohm's wheel illustration by putting your thumb over the value you are looking for. The visible values will tell you what mathematical operations to perform to get your desired results.

For instance, to calculate the power output of a solar panel, multiply the voltage output of the panel times the current output (amperes) of the panel. This gives you the power output in watts.

To calculate power output or input, and you know the current and the voltage: **P = IE**

To calculate how much power (watts) a specific appliance uses, look for the voltage rating and ampere rating on the appliance and multiply them to obtain the watts used. These are important figures for sizing a power system so that the daily power output matches or exceeds your total daily consumption.

At right is a more extensive version of the wheel with power formulas.

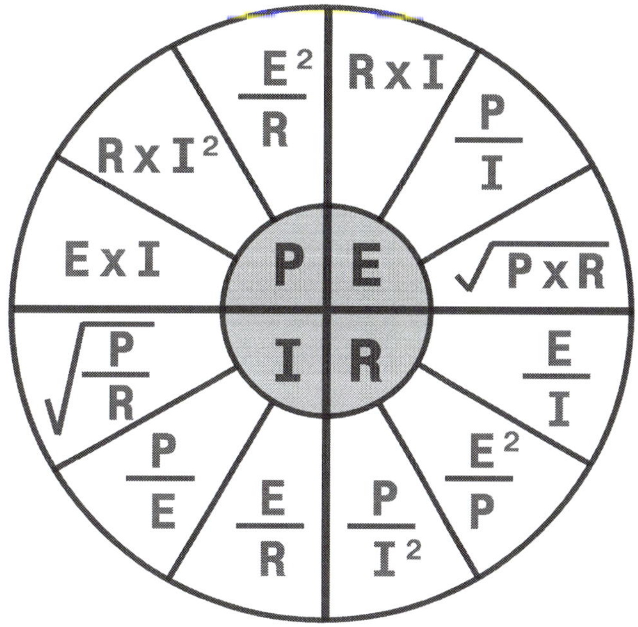

Ohm's wheel with power formulas

Measuring instruments

To measure current, voltage and resistance in photovoltaic power system circuits, a digital multimeter is generally used. It is a good idea to have two to three inexpensive multimeters on hand – or one really good one. The accuracy of inexpensive multimeters can be questionable. There is really no substitute for a very good expensive multimeter. For most testing, it may not be that important; but for calibrating charge controllers and other components more accuracy is needed.

The most frequently used multimeter functions for photovoltaic power system construction are voltage, current, resistance, and continuity.

Continuity indicates whether there is a conductive path which will allow electricity to flow between two points. A continuity test of circuits that you have assembled, or components that you have connected, will show if there are problems with shorts, bad connections; or if everything will conduct to the points as you expect.

Most multimeters also come with a diode tester; some have a transistor check, a capacitor check and a variety of other functions. Some have software so that you can connect the meter to a computer and log data over a period of time. This can be handy to monitor a newly installed system and observe performance in case any tweaking is necessary; or to obtain and store your system's performance data for other reasons.

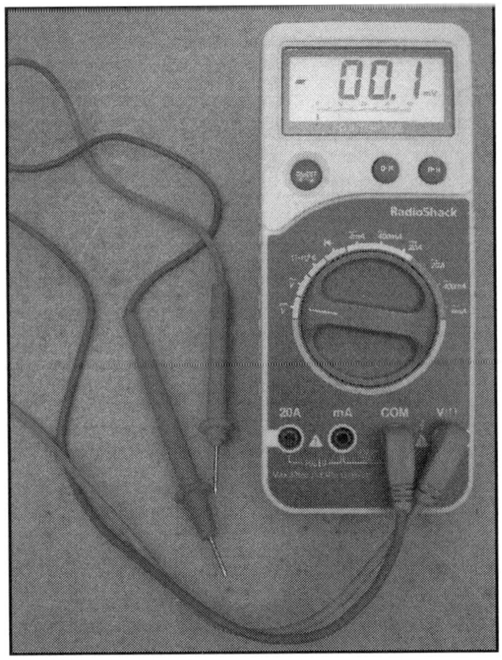

Digital multimeter

A multimeter for your photovoltaic system should be capable of measuring up to 20 amps. You can use separate meters for voltage, current and other readings, however the convenience of a multimeter is usually preferable.

Designing a photovoltaic system

Photovoltaic systems are sized by assessing your power needs and budget. If you have large power needs and a small budget, you will have to make adjustments. If budget is not a concern, then you can have what you want. Most of us, however, have to balance these two factors before purchasing or building panels and other system components.

The size of a system is usually designated in terms of the number of panels in the array and their output. For instance, an array of ten 100 watt panels will be designated a 1000 watt system.

- Any array up to 1000 watts is a "small" system.
- From 1000 watts to 5000 watts is a "medium" system.
- Anything from 5000 watts on up is a "large" system.

Calculating energy needs

Figuring out your energy needs can be very informative. Every household and each one of us has an energy use profile. Personal lifestyle has everything to do with energy consumption. How much time and what time of day each household member spends awake and active at home, and what they are doing during that time all factor in to total electrical usage, as well as the actual appliances used.

Once you have some idea of what appliances you wish to run on your PV system, there are many ways to calculate system needs, some quite complex and others simple and straight forward. Several free software programs are available that give fairly sound technical analysis as well as evaluations of cost and payback for your designed system. There are some good text books with sound sizing information and insolation tables. And, a little internet research will provide plenty of information about system sizing. Some URLs and books are listed in the resources section starting on page 179. For a ballpark estimate, I have included some sizing methods that work well for small PV systems.

Watts per square meter

Most estimation methods use the unit *watts per square meter* (W/m^2) for calculating the output of a solar module. Output of panels is tested under an irradiance value of 1000 W/m^2 (which nearly coincides with terrestrial energy received from the sun), when the surface of the panel is directly facing the sun, and the sun is more or less directly overhead. This convention, although not exactly perfect, at least provides a baseline for a general evaluation of module performance in the field.

Insolation tables and maps

Insolation tables and maps show the number of peak hours (kWh/m^2) on a daily, weekly, or monthly basis at a given longitude and latitude and time of year. This number of peak hours is used to calculate the output of panels at a particular geographic location in a given time frame.

Many insolation tables and maps only provide figures for horizontal flat plate collectors, which is not adequate information for sizing photovoltaic systems unless you are living directly at the equator. If you don't live at the equator you will have to use an error correction formula for the angle of the panels to get the correct figure from the horizontal panel data. The US Solar Radiation Resource Maps has a very good selection of tilt angles to input into your calculations and they are easy to use.

Maximum daily usage

Make a list of the electronic and electrical appliances you wish to power with the PV system on a regular basis. Most appliances are labeled with the voltage and amps needed to operate the device. Multiply volts times amps to get the watts. Some appliances have phantom loads, meaning that even though the appliances are not operating they are still using current. VCRs, TVs and other appliances (especially anything with a remote control), need to be checked for their phantom power drain and calculated into your figures.

A watt meter accurately measures any appliance's electrical consumption.

Also, some appliances (for instance refrigerators and water pumps) cycle on and off throughout the day, so you may have to get the average run time per day and power usage in watts from the manufacturer. However, the best way to ascertain the daily power requirements of appliances such as refrigerators and devices with phantom loads is to use a meter such as a watt meter. This is available from most solar suppliers and other sources on the web. All you need to do is to plug the watt meter into the wall socket or power strip, then plug an appliance into the watt meter. It will tell you at a glance what the appliance's daily watt usage is, and much more.

System sizing Method 1

The first sizing method is based on calculating what you want for output – the power demands of all the devices you wish to run on the PV system – and sizing the system components to meet those demands.

Maximum demand

See the table on the next page. For each appliance, mark down the wattage. Next to the wattage, mark the total time in hours that the appliance is on for each day of the week, and multiply the watts times the hours to get the daily power usage of the appliance. Total the watts for each day, then choose the day with the most watts. This is your maximum watt hours used per day, and indicates what size solar array and battery storage bank you will need.

Next, list all the essential "must function" appliances that are always on, and will draw power whenever they need to run, such as refrigerators and water pumps. Then make a list of all the other appliances. From this list, make smaller lists of appliances that you wish to be able to have running at the same time. Choose the group with the most watts, and then add the watts of this group to the total wattage of the "must run" appliances. This is the maximum watt hours that will be needed at any one time, which is the figure to use to choose the size inverter you will need, if you will be using an inverter.

DAILY POWER USAGE

APPLIANCE	wattage	SUNDAY		MONDAY		TUESDAY		WEDNES-DAY		THURSDAY		FRIDAY		SATURDAY	
		hours	watts	hours	watts	hours	watts	hours	watts	hours	watts	hours	watts	hours	watts
TOTAL DAILY WATTS															

An interactive PDF version of this table is available to download at www.goodideacreative.com/solarii.html

Daily insolation and panel output

Go to the solar radiation resource maps (see page 179) appropriate for your area and find the average amount of incoming solar radiation for your location, which is expressed in terms of KWh/m^2/day. Find the month with the lowest insolation number. This is the month which generally has the least amount of sunlight during the year at your location. Let's say the daily insolation number is 3.5 for December, and that's the lowest value for the year.

To figure the daily output of one PV panel, multiply the short circuit current times the daily insolation figure. If the short circuit current is 4 amps and the daily insolation figure is 3.5, then daily output would be 14 amps. Multiply this figure by the system voltage, let's say 12 volts. This gives a final figure of 168 watts, the daily output of one of your panels, and since you're using the lowest daily insolation figure, this is the low end of what you can expect from the panel on a daily basis.

(daily insolation x short circuit current) x system voltage = daily output

Next, calculate the number of PV panels needed to adequately charge your batteries to supply sufficient watts to power your appliances. Let's say your maximum daily requirement is 810 watt hours. Divide 810 watts by the daily output of one panel as calculated above (watt hours per day at 12 volts, in this case 168 watts), which equals 4.82. Then, divide this number by the charging efficiency of your battery, which is expressed as a percentage. You can either use the battery manufacturer's specifications, or use a ballpark figure of 80% for lead acid batteries and 70% for nickel cadmium. So, for lead acid batteries, divide 4.82 by .80 (80%), which is 6.025. This indicates that six panels are needed.

maximum daily usage / daily output = y
y / charging efficiency of batteries = number of panels needed

Sizing the battery array

Every location has its days off, when the sun just does not shine, and your batteries do not get charged at all. So, you have to figure how many days of no charging that you need to get from your battery bank. For a remote location that is unattended and it is critical that power is always available from the batteries, a ballpark figure would be from 7 to 10 days extra capacity. In an area that is frequently cloudy, consider five to seven days of extra battery capacity. Try to size the batteries for at least five days extra capacity – unless you live in a generally sunny area, in which case 3 to 4 days reserve is probably sufficient.

The extra days needed are what is called your usable capacity requirement. Multiply your daily watt hour needs (in this case 810), by the number of storage days required (for instance 5 days) which is 4050. Divide 4050 by 12 volts which gives 337.5 – the total capacity needed in amp hours for the battery array.

maximum daily usage x storage days = y
y / system voltage = amp hour capacity for battery array

With your requirement for the total usable capacity figured out, you can calculate the number of batteries that you will need.

Divide the total usable capacity needed (337.5 amp hours at 12 volts) by the full capacity specified for the battery you are using.

Let's say you are using Trojan 105 6 volt batteries with an amp hour capacity of 225 amps. Divide 337.5 by 225 which is 1.5. Then divide 1.5 by the maximum depth of cycle allowed , which in this case would be 50% (0.5). The result is 3, which indicates a minimum of three 12 volt batteries, but since we are using 6 volt batteries, six Trojan 105 6 volt batteries connected in series/parallel are the minumum needed.

amp hour capacity for battery array / battery full capacity = y
y / max. depth of cycle = min. number of batteries

System Sizing Method 2

Another approach for sizing a system is to start from the budget you have for the system, or system components you may already have. Size the system accordingly and then figure out what devices you will be able to run with it.

The amp hour method is a simple way to size small PV systems, and is the method I used for planning the Solar II system. This system was intended to be used as a daily power source for some small appliances (lights, computer, TV), and as backup during power outages to run a few low demand appliances.

I didn't have much to spend on the project, but I figured I could afford about four PV panels if I made them myself; and two to four 6 volt batteries.

First, I checked the insolation tables for my longitude and latitude and found the month of the year with the lowest hours for daily insolation. At my location in Vermont, that is December, with an average of 3.5 hours with a south facing panel tilt of my latitude +15° (60°). So, four PV panels with an output of 3 amps each, connected in parallel (4 panels x 3 amps = 12 amps per hour), would put an average of 42 amps a day into the battery bank in the least sunny part of the year.

3.5 hours x 12 amps per hour = 42 amps
daily insolation hours x total array output = daily amps going into the batteries

This is an important figure as it tells me how much I can use every day. If I use 60 amps a day I am going to run into a deficit and that is not going to work. In reality I can not even draw 42 amps because it takes 10% more energy to recharge the bank each time it is charged, and I also need to derate this input by at least another 10% to account for system losses and shortfalls. That means that I will be able to draw about 33.6 amps per day if my panels are supplying 42 amps per day to the batteries.

At this point I assessed the appliances I wanted to run and calculated how many and for how long they can be used. If the appliances add up to a draw of about 10 amps per hour, then they can be run for a little more than three hours.

Next is to figure out how many batteries are needed. Since I am using Trojan T105 6 volt batteries, at least two connected in series are needed to add the voltage for a 12 volt system. Two batteries connected in series will give the same amp hours as one battery. That would be 191.25 amp hours for the Trojan T105s.

The T105s are rated by the manufacturer at 225 amp hours but they need to be derated by 15%, which gives the derated amp hours of 191.25. It's a good idea to not discharge batteries more than 30% if you want to get maximum life out of them. So, drawing about 57 amps from the battery a day should keep them above the 30% range (191.25 x 0.3 = 57.38). However we're already limited to a withdrawal of 33.6 amps by the capacity of the PV panels, so two 6 volt Trojan batteries connected in series should be adequate.

If AC appliances are going to be used, an inverter is needed. It is good practice to match the inverter output rating as closely as possible to the load. In this case, sizing the inverter is easy. The system will only be running a few small appliances at a time such as one or two energy efficient fluorescent lights, and a desk-top computer or small TV. These combinations of appliances will take about 125 watts.

In the calculations above I found that I could use 33.6 amps at 12 volts, which would give me about 3 hours of run time per day. I multiply 33.6 amps x 12 volts to determine that there are 403.2 watts available from the batteries according to my usage parameters. Since inverters are rated in watts, this is the figure used to size the inverter.

allowable amps per day x system voltage = daily watts available from batteries

The inverter will use up 5% to 15% of the energy available when converting DC to AC, so I will assume that 15% of the 403.2 watts will be used up in conversion so I have about 342.7 watts available. This will give a little over 3 hours run time as indicated from the earlier calculations. However, what is more important is that I will only be using about 125 watts so I do not need a very large inverter. Inverters are more efficient when the load is from 30% to 90% of their rating. If I choose a 300 watt inverter, the 30% to 90% range would fall between 90 watts and 270 watts. Since I will be using about 125 watts per hour, this is a decent match. This configuration with two 6 volt batteries seemed fine until I took into account that there are frequently many cloudy days in a row in my location, especially in the month of lowest daily insolation. If the battery bank is discharged once, I would have to wait until it recharges before drawing any more amps out to stay within the batteries' charge capacity.

Since this a small daily use and backup system I'm not too worried about it because it is supplemental, and not critical. What could happen, however, is the batteries are not recharged because of cloudy weather, and then we have a power failure on the grid. Since I am concerned about this scenario, I chose to add two more batteries to the array. Each pair is connected in series within the pair, and then each pair is connected to the other pair in parallel. This gives me twice the amp hours. I will have to make up the amount of power I draw, but then again I can switch to grid power if there are too many days in a row with no sun. If you are not on the grid and have a small PV system, then you have to monitor your usage and perhaps use a small back up generator to charge the batteries occasionally to get through those long cloudy spells.

I used the months with the least daily insolation because I am most concerned about system output in the winter months. At my location, during the winter there are fewer hours of sunlight, and I generally use more electricity.

Other considerations for sizing a PV system

In some situations, space requirements for the panels may be a concern, for instance in urban areas. If space is limited, you will want PV panels with the most efficient solar cells possible, and the space you have available for PV panels may dictate the size of your PV system above all other factors.

The efficiency of the cells in a panel determine how much power can be generated in a given space. The larger panel in the photo is rated at 180 watts. The smaller panel has more efficient cells and is rated at 200 watts

To assess the area needed for PV panels, first consult the daily insolation (incoming solar radiation) charts for your geographical area. You can use either the figure for the month with the lowest or highest daily average, or the average figure for the whole year, depending on your needs and use patterns. For illustration I am using 3.5 hours at 1 kw per square meter, the daily average of my location's month with the least sun, so my system calculations will represent the low end of performance that I can expect. Most solar cells have an efficiency rating of 12% to 15%. This means that every square meter of panels should produce about 150 watts. Multiply 3.5 hrs. times 150 watts, which equals 525 watts per square meter per day. Then, for power needs of about 10 kw (10000 watts) of energy each day, divide 10,000 watts by 525 watts, which equals about 19 square meters.

If I divide the 10,000 per watt day electrical needs by 3.5 hours, I can see that I would need to produce 2,857 kilowatts per hour to fulfill my needs, and the panels will cover a space of about 19 square meters (204 square feet), or an area a little larger than 10 feet by 20 feet.

AC or DC

At the latter part of the nineteenth century there was a growing need for widely available and cost effective electrical power for an emerging industrial and consumer economy. In response to this need, Nikola Tesla invented the high voltage AC generator, which could supply electricity at long distances from the power generating source. AC electricity, when stepped up with a transformer to higher voltage, travels longer distances with much less loss than DC current does. This model of power production is called CPG or centralized power generation, with large power plants generating electricity which is distributed to many consumers. At the time, this was called the Westinghouse system. Another CPG system vying for commercial acceptance was the Edison DC system. Unfortunately the DC system wasn't suitable for long distance transmission – it required much thicker wires to offset the voltage drop encountered in the system. Edison generating plants could handle several square city blocks of power distribution adequately, but could not compete when it came to the many miles of distribution necessary for a really centralized system. This led to the adoption of the Tesla AC system, which was a good choice for the times.

This has had quite a profound effect on the consumer generated power (CGP) industry. The greatest effect has been in appliance availability and cost.

Since the widely adopted model was AC, appliances were manufactured to operate from AC current, and manufacturers quickly dropped their DC models.

Some DC advocates claim that DC is a much better way to go than AC for CGP because it is more efficient and costs less to set up and maintain. The problem is that at this point in history, AC products are much more widely available, and lower cost than DC appliances because they are produced in such quantities.

In the emerging alternative energy market of the 1970s, some manufacturers tried to address this deficiency by supplying some DC products to consumers, but in the long run, they could not break through the price point barrier that was necessary for these businesses to survive. There was not enough interest or need on the part of consumers to make the change.

Even though DC power production is more efficient for CGP with alternative energy systems, there is not much enthusiasm for its use. So, a whole industry has grown up around providing AC from DC sources, and in fact has positioned and marketed itself as a central component to any alternative energy enterprise.

This industry is the inverter market. Inverters modify DC current to AC current. The downside of inverters is that they use up energy in the conversion, and so are not as efficient as direct use of DC from a photovoltaic system.

There are also growing concerns about possible malignant effects of AC electrical fields on the human body. If this is enough of a concern and you are willing to pay more to get DC appliances, you can set up your PV system to be strictly DC.

That being said, inverters are handy to have because most of us already have AC appliances. Also, 120 volts AC can travel further without large voltage drops, so if you have to have long wire runs in a system, lighter gauge wire can be used. Without an inverter, DC on long wire runs requires heavier gauge, more costly wire to compensate for voltage drop.

Still, what's on the end of the wire will probably have more to do with energy loss than the wire size, which is why it's so important to take the time to carefully assess the appliances you intend to use. For instance, there are DC refrigerators that are many times more efficient than conventional AC refrigerators. Good places to look for DC appliances are vendors for RV, marine and, of course, alternative energy markets.

Conservation

Obviously, it can be quite costly to purchase and set up a photovoltaic system. When most people do the calculations to size a PV system to match their household electrical usage, sticker shock sets in. If you build your own system, you can reduce the cost as much as half. But, reducing your electrical demand, and thus the size of the PV system you need, should be a serious consideration in any case.

Energy conservation is a significant part of designing, setting up and using a PV system. The fact is that most of us, when hooked up to the grid, are not very aware of the details of our energy consumption; and the manufacturers of the electrical appliances we use have mostly not cared about making them energy efficient. This is changing (though not fast enough for some of us!), and new technologies and materials are making it possible for appliances to perform their tasks with much lower energy usage. Still, don't assume that all new appliances are energy efficient, and don't even assume that you can rely on efficiency ratings to tell you the whole story.

Your "household ecology" – how you use an appliance and in what context – plays a big role in whether or not a given appliance is the most efficient model for you. Case in point: a washing machine will get points toward being rated energy efficient if it runs a longer spin cycle because it gets more water out of the clothes, and thus reduces the time the clothes dryer has to run a heating element to dry the clothes. However, if you line dry your laundry and thus don't use a dryer, the longer spin feature of the washing machine is going to increase instead of reduce your electrical consumption.

Another example of household ecology: where is a refrigerator or freezer located? If it sits right beside a source of heat, for instance, the kitchen stove, it is going to run more and use more electricity than if it is in a cooler location.

Lighting is probably the easiest place to implement energy conservation, if you haven't already. Compared to incandescent light bulbs, compact fluorescent bulbs are very energy efficient. The fluorescent bulb in the photo gives the light equivalent of a 60 watt incandescent bulb but only consumes 14 watts.

Energy efficient compact fluorescent light bulb

Halogen lights are fairly efficient, but the bulbs burn hot and can be a fire hazard if not used correctly. LED light bulbs are even more efficient than flourescents. They have extremely low operating temperatures, last a very long time, and the technology is maturing quickly, so the quality of LED light is improving greatly. LED lighting will eventually replace most major forms of lighting for general consumer use. The LED bulb in the photo puts out the equivalent light of a 50 watt incandescent bulb and consumes 12.7 watts.

LED light bulb

As you become your own power company, you will want to also become a diligent investigator of energy efficiency. Ask lots of questions, get the details, make lots of comparisons, and ponder the way the device will be used before you purchase anything you plan to plug into your PV system.

More system choices

Solar electric systems can be DC only, DC-AC, or AC only.

DC only systems are very cost effective and initial capital outlay for the system components is less, though it can be expensive if you have to replace a lot of AC appliances with DC models.

DC-AC systems can be configured with dedicated DC lines to specific DC appliances such as refrigerators, pumps, etc.; and inverted power for other appliances. They can be cost effective especially when output is matched to the appliance used.

AC only systems are cost effective in certain circumstances but in general are the most costly in terms of initial outlay. The advantage of the AC only system is readily available wiring, outlets, fixtures and appliances.

For a small system consisting of a few solar panels to experiment with, there is flexibility to change modes, so it's not necessary to commit to AC, DC or AC-DC. But for systems installed to power a home totally, homework is needed to figure out what will work best for you. There is a great deal of material on the web comparing types of systems and energy efficiencies of different appliances.

Hybrid Systems

There are certain deficiencies inherent in a purely photovoltaic system, for instance, in some locations at certain times of the year, there may be many days of no or very little sun. Fortunately, photovoltaic systems work well in combination with other renewable energy sources such as hydrogen fuel cell systems, wind, hydro, geothermal, and tidal. For instance, in my area at the times of the year when there are more cloudy days than not, the wind is usually blowing. Rather than rely on a fossil fuel generator or the grid to supply the needed power during these times, a wind generator, or some other renewable energy source can be incorporated into a PV system to take up the slack.

Other combinations are possible such as adding a gasoline, propane, or diesel generator, which reduces the size of the PV array needed to get through the worst weather conditions. The generator can be run when no sun is available and can charge the batteries as well as supplying power directly for short periods of time.

It is beyond the scope of this project to go into the details of incorporating other energy sources into a PV system, but I will briefly discuss using hydrogen production in PV systems to use solar electricity that would otherwise be wasted.

Solar, hydrogen and fuel cells

Hydrogen production system powered by PV panels

During the peak insolation periods of the year, most solar panels charge the batteries to capacity very quickly and then shunt that electricity to no use in most systems. With a hydrogen generating electrolyzer in the system, instead of shunting and wasting the excess electricity, that power is used to produce hydrogen, which can be stored indefinitely. The hydrogen can then be used when very little sunlight is available and the battery banks are not getting fully charged.

Fuel cell generator

At those times, the hydrogen is fed into a fuel cell stack. The fuel cell stack converts the hydrogen to electricity, heat and water. Some of the advantages of a hydrogen photovoltaic hybrid system are: hydrogen stores indefinitely, it can be an energy carrier for electricity that would otherwise be wasted, and the end result is pollution free.

Batteries do not have the carrying time that hydrogen does. Batteries have to be constantly cycled, charged, and discharged, and they discharge very quickly in use. If the weather does not agree with your batteries' time frame – which usually happens in the winter – you will need something for backup. If you don't have access to, or prefer not to use power from the grid, and want to generate the electricity you need without creating air pollution, you may want to consider incorporating hydrogen in your PV system. Hydrogen hybrid systems are still in the experimental stage, but it's quite feasible to build your own to see what they are all about.

Of course, you could dispense with the batteries altogether and simply use a photovoltaic hydrogen fuel cell system year round. This would eliminate having toxic lead batteries around and dealing with disposing of them properly at the end of their useful life.

Grid tied solar

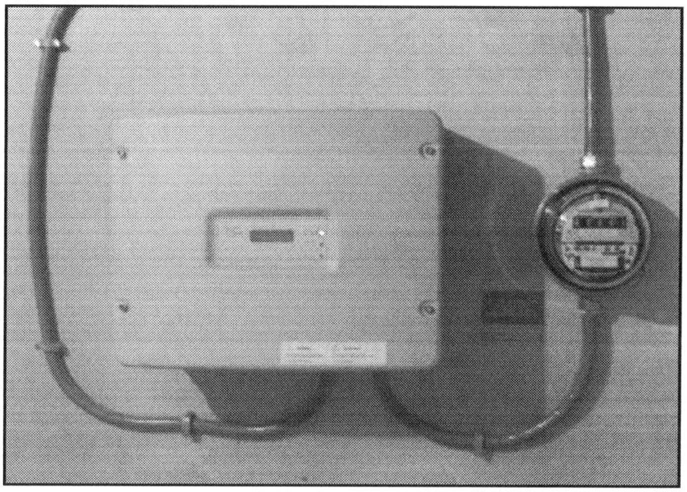

photo courtesy Independent Power & Light

Grid intertie with net metering

Another system option is to have it connected to the grid, and automatically switch to utility power when the batteries get low and aren't getting charged, and then switch back to the batteries when they are charged again. This can be a very cost effective way to go for certain situations. The important thing to remember is that utilities have a minimum monthly charge whether you use their electricity or not.

Net metering

Then, there is utility intertie with net metering. With this scheme your PV system output goes into the grid, you draw your electricity from the grid and the utility company may pay you if your system puts more power into the grid than you used. (However, you still have that monthly connection fee.) You can have a battery bank or not. If not, the grid essentially becomes your only power storage, which means that if the grid goes down, you have no electricity.

The scheme sounds good, but be sure to do the research and math if you want to anticipate the real costs and benefits of net metering. This is new territory for most utilities and regulators of utilities, so standards are still being worked out and vary widely. Find out what your AHJ (authority having jurisdiction) and utility require.

Some of the questions to ask your utility are:

- What is required for liability insurance?
- What is required for hardware specifications, installer certification and inspection?
- What is the hook-up fee?
- What monthly fees will you be charged (ie, connection fee, administrative fee, etc.)?
- At what rate will you be compensated for the power you put into the grid?
- Is there a cap on how much power you will be compensated for, and is excess carried over from one billing period to the next?
- Other hidden costs?

It may not work out as well as you may have expected. If you're thinking about a system with a grid intertie and net metering, be sure to do your homework.

Electronic Stonehenge - Siting a Solar Array

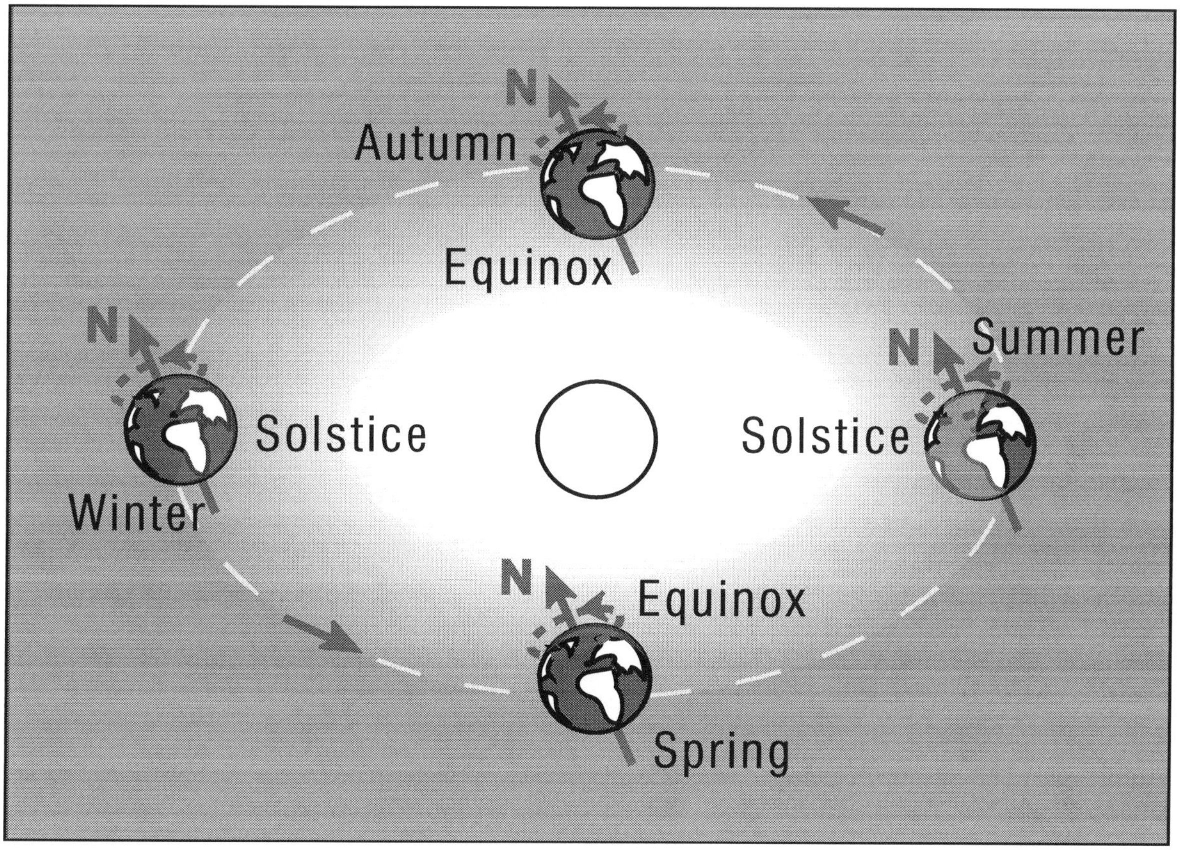

The tilt of the Earth's axis causes the changing seasons.

Our planet orbits the star we call the Sun at a breathtaking speed of nearly 67000 miles an hour. It also rotates on its axis at nearly 1000 miles an hour. Although we never give it much thought, these movements greatly affect what we do on our planet on a daily and seasonal basis. The daily rotation of the Earth gives the appearance of a rising and setting sun. It is a powerful timer for human activities.

The Earth's orbital motion (which covers a distance of 584 million miles) and the 23.5° tilt of its axis creates the seasons for us.

The ancients built stone structures and temples oriented to mark the exact times of certain points in these cycles. For them it was a matter of harvesting photons from the sun for agricultural purposes. For us PV enthusiasts, it is a matter of harvesting photons for direct electrical power generation.

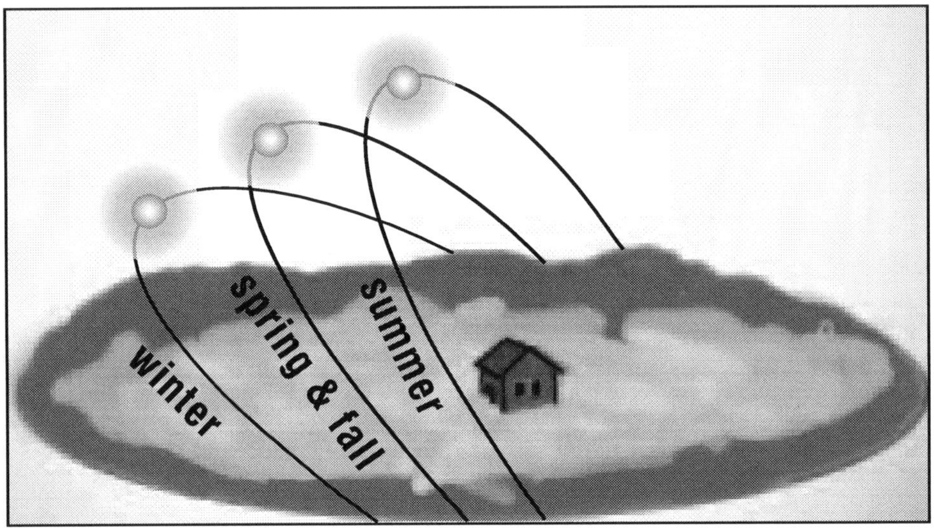

Solar azimuth

Each of us experiences the Sun in a way unique to our particular location on the Earth. For instance, where I live in a temperate zone at about 45° north latitude, the sun rides high in the sky during the summer months, so I get long days and thus more light. In the winter the sun rides very low in the sky, giving me shorter days and less light. On the equator you would experience days of equal length throughout the whole year. At either the Arctic or Antarctic Circle there would be at least one day in the summer when the sun does not set, and one day in the winter when the sun does not rise.

Solar array in winter position

Positioning solar panels

If you live north of the equator, your solar panels should face true south. South of the equator, the panels should face true north. If you live on or very near the equator, face your panels upward.

In addition to orienting PV panels to true south or north, they should be angled directly toward the sun to get the maximum out of the photon harvest. If the panel angle will not be adjusted and thus be one permanent angle throughout the year, the tilt angle should be the latitude of the site. For instance, at 25° north of the equator, the panels would be tilted at 25° facing south.

You will get better performance from your panels if you adjust the tilt for each season. At latitudes more than 30° either north or south of the equator, set your panels at latitude +15° for winter, and latitude -15° for summer. During the spring or fall, the angle will be the same as your latitude. For example, where I live in Vermont, at 45° north latitude I set my panels at 60° in the winter and 30° in the summer. In the spring or fall, they are set to 45°.

What I usually do is change the angles at the midpoints between the solstices and equinoxes. So, the schedule I use is:

- 60° (latitude +15°) on November 7th
- 45° (latitude) on February 5th
- 30° (latitude -15°) on May 5th
- 45° (latitude) on August 7th

In the southern hemisphere you would switch the May and November angles. For latitudes between 15° south and 15° north, a tilt angle of 15° is used. Even if you are on the equator, the panels should be angled so that rain will run off.

Although you can set the tilt angle once for the degree of your latitude and leave it that way, adjusting the angles seasonally will give roughly a 10 percent increase of solar energy reaching the panels, which is a significant boost to power production.

Solar trackers

Mechanical trackers turn a PV array to follow the sun. This can be beneficial in some circumstances, but they may not provide enough extra power in the winter to justify the cost of the tracking system in a location with frequent cloudy conditions.

Tracking systems can be either single axis or dual axis. Most single axis systems follow the course of the sun from the east to the west, and then reset themselves to the east for the next sunrise. The dual axis does the same thing, but also adjusts to the sun's angle from the horizon.

Generally, in the north and south temperate zones more energy is needed in the winter, and although the trackers provide a great increase in energy output in the summer, they do not provide nearly as much of an increase in energy output in the winter. In temperate zones, panels with trackers can receive 10% more energy in winter and up to 40% more in the summer than fixed panels, according to some sources. Of course, any energy boost in the winter is welcome, but the added cost of the tracking system should be weighed against the benefits to decide whether it is a good investment.

Tracking systems can be especially advantageous to use in solar hydrogen systems where the excess electricity that is generated in the summer can be used to make hydrogen, which can be stored and used in the winter to generate electricity in fuel cells.

Choose a shade free site

The first consideration for siting a solar array is an unobstructed, shade free, south facing location (or north facing if you live south of the equator). If you live on or near the equator, then a generally unobstructed shade free site will do, without concern for direction

Avoid siting solar panels in areas that do not get full sun all day if at all possible. Panels should face true south or north without any partial or full shadows. If a portion of any panel is shaded at a certain time each day, be sure to have a bypass diode installed with each panel, so that the performance of the shaded cells does not bring down the performance of the whole array.

Wire runs

Solar panels arrays should be located as close as possible to where the electrical output will be used to reduce the cost of the wire run, and to reduce voltage drop due to long wire runs. The ideal length of wire run from the panels to the battery bank is no more than 100 feet. The best distance for a small 12 volt system would be 50 feet or less.

Prepare the site

When you have chosen an appropriate site, clear out any obstacles and high grass. The area should be mowed and leveled as much as possible for accurate layout and construction.

Site orientation

One way to find true north or south is the shadow method. This is most accurate when performed on a sunny day on or near the autumn or spring equinox. The nearer you are to the equator, the better the accuracy of this method. All you need is a straight stick (a dowel rod, for instance) about 36" long, some string, a tape measure, and two stones.

Several hours before midday, insert the rod into the ground making sure that it is straight. You can check it with a plumb bob to insure accuracy. Notice that the stick produces a shadow. Measure the length of the shadow with a tape measure, or a length of string and mark it where the shadow ends. Mark the tip of the shadow with one of the stones or some other object.

Several hours after midday, the shadow of the stick will have changed position. When the shadow is the same length as the first shadow length measurement you took, place a stone or other marker at the tip of the second shadow. Run a piece of string between the two markers, and this straight line is the east-west line. From the center of the east-west line, looking toward the stick is facing true south if you are north of the equator; or true north if you are south of the equator. This is the reference for laying any foundation or blocks and for orienting your panels to true north or south.

Using the shadow method to get an east-west line.

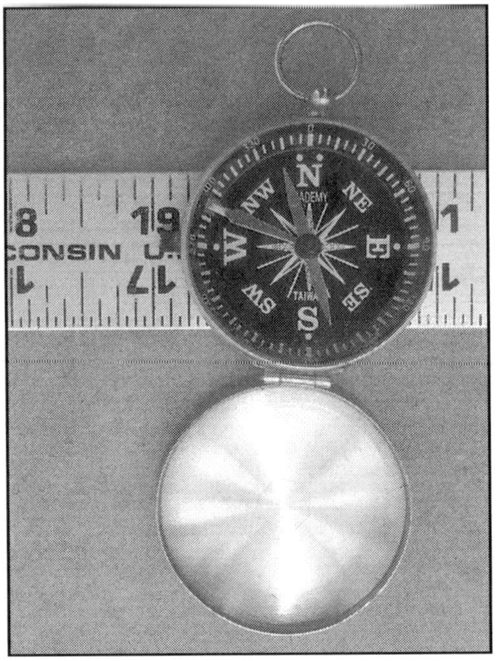

Using a compass and magnetic declination correction to find true north.

Another way to find true north or south is to use a compass to find magnetic north and south; then make the correction for your location to get true north or south. This correction to compass readings is called the magnetic declination. The magnetic declination for your specific area can be found at the National Geophysical Data Center, see page 180 for the URL.

This method has the advantage over the first in that you can do it any time of the year, sunny day or not. It also has a higher degree of accuracy. For instance, where I live at 44° 54′ North Latitude and 72° West Longitude my magnetic declination and thus the correction for true north is 15° 39′.

Photovoltaic System Components

A complete stationary PV power system usually includes:

- PV panels
- Combiner box
- Circuit lightning arrester
- Panel rack and mount
- Grounding system for panels, racks and mounts
- Batteries
- DC service box with charge controller, cut off switches, and/or fuse
- Inverter
- AC service box with circuit breakers
- DC outlets and switches, or AC outlets and switches
- Wire for:
    - Solar panels to combiner
    - Combiner to service box
    - Service box to battery
    - Battery to inverter
    - Inverter to AC outlets and switches
    - DC outlets and switches

Parallel and series circuits

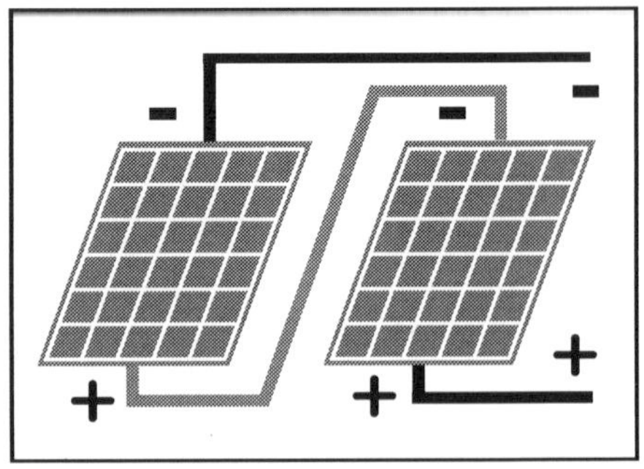

Two panels connected together in series

In order to design a PV system, a basic understanding of parallel and series circuits is needed to configure voltage and current outputs from the photovoltaic panels and the batteries. Connecting panels in series adds voltage, connecting panels in parallel adds current. Most photovoltaic systems are configured to be 12 volt, 24 volt or 48 volt.

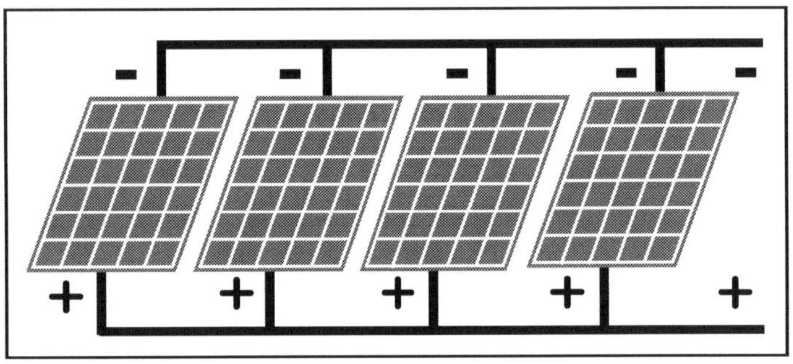

Four panels connected together in parallel

For a 12 volt system with more than one panel, connect the panels in parallel. For example, four 12 volt panels that each put out 4 amperes at about 18 volts, connected in parallel would have a final output of 18 volts and 16 amperes.

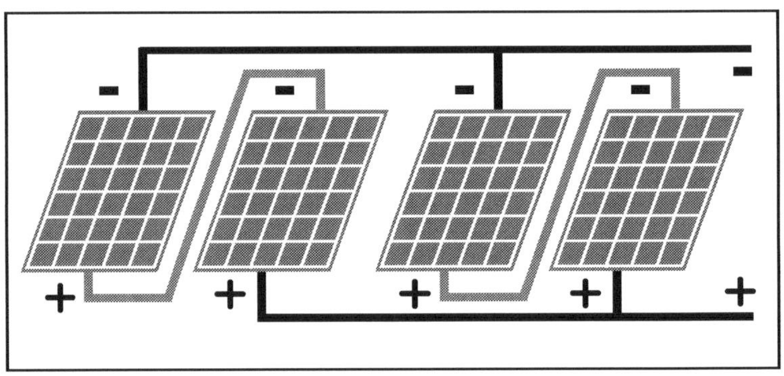

Two pairs of series connected panels connected to each other in parallel

For a 24 volt system, connect two 12 volt panels in series; but, if you have 24 volt panels, they should be connected in parallel. If you have more than two 12 volt panels, use series connections to make pairs of panels, and then connect the pairs together in parallel.

Battery banks are configured in the same way. For a 12 volt system using 6 volt batteries, connect two batteries in series. If each battery produces 6 volts and is rated at 225 amps, when they are connected in series you will have a 12 volt power source at 225 amps. The current remains the same but the voltage is added. If you connect the same two 6 volt 225 amp hour batteries in parallel, you would have a 6 volt 450 amp battery supply.

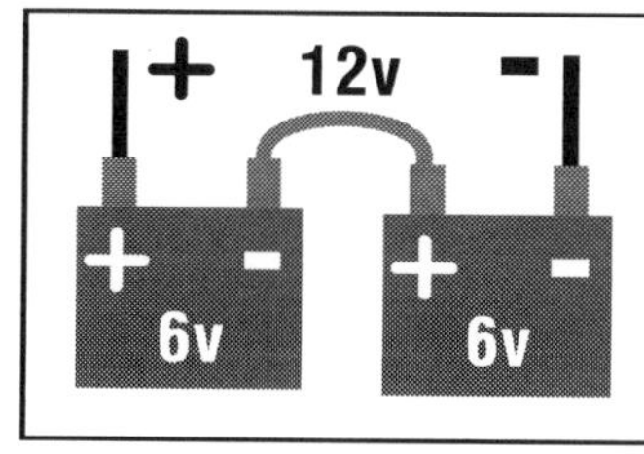

Batteries connected in series

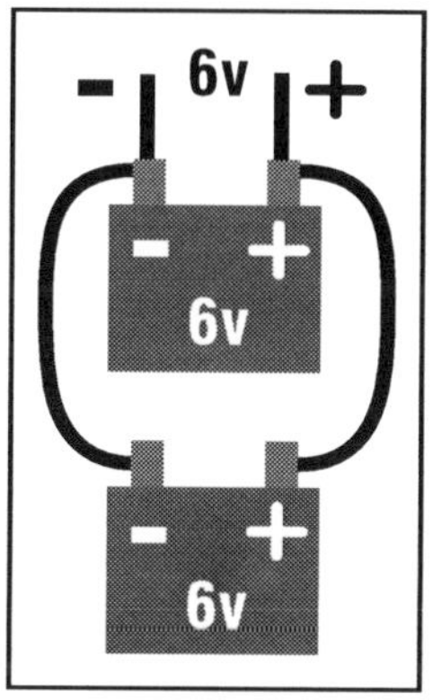

Batteries connected in parallel

If you series connect two 6 volt, 225 amp hour batteries to make a pair, and then parallel connect the pair with another 6 volt, 225 amp hour series connected pair, you would have a 12 volt 450 amp battery supply.

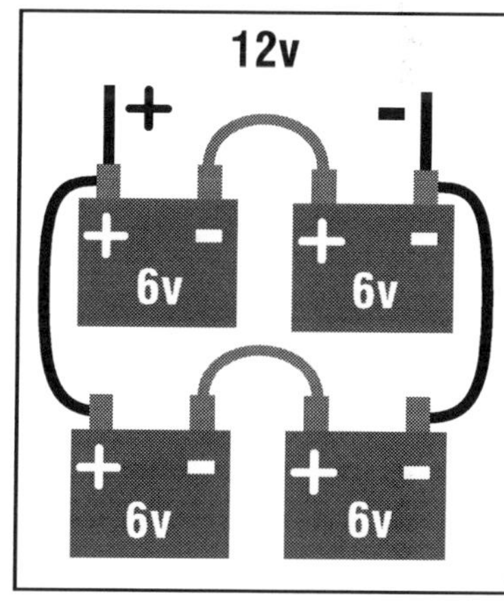

Two pairs of series connected batteries connected to each other in parallel

Customizing output

As noted in *Build Your Own Solar Panel*, photovoltaic panels can be fabricated to produce the standard 12, 24, or 48 volt output, or they can be designed to any specification to exactly match your needs. 12 volt panels are the standard panel used for most small PV power systems. 24 volt panels are used in medium sized systems or systems where the wire run is long. 48 volt panels are used for long wire runs and for large PV power systems. Commercial 12 and 24 volt panels are readily available from most PV suppliers.

12 volt panel arrays can easily be configured to output 12, 24, or 48 volts:

- Two 12 volt, 4 amp panels connected in parallel to get 12 volts at 8 amps,
- Two 12 volt, 4 amp panels connected in series to get 24 volts at 4 amps,
- Four 12 volt 4 amp panels connected in series to get 48 volts at 4 amps.

Similarly, 24 volt panels can be configured to output 24 or 48 volts:

- Two 24 volt, 7 amp panels connected in series to get 48 volts at 7 amps,
- Two 24 volt, 7 amp panels connected in parallel to get 24 volts at 14 amps.

Panels can also be configured in combinations of series and parallel connections:

- Two 12 volt, 4 amp panels connected in series can be parallel connected to two other 12 volt, 4 amp panels connected in series for a total output of 24 volts 8 amps.

12 volt is the standard for small systems and for systems that have short runs of up to 100 feet. 24 volt systems are standard for longer runs of up to around 300 feet and/or if you simply want a 24 volt system. 48 volt is the standard for longer wire runs of up to 400 feet. Of course you can use any voltage you wish for short runs. The advantage of higher voltage is that it reduces voltage drop and so is a better choice for long wire runs.

Wire in Photovoltaic Systems

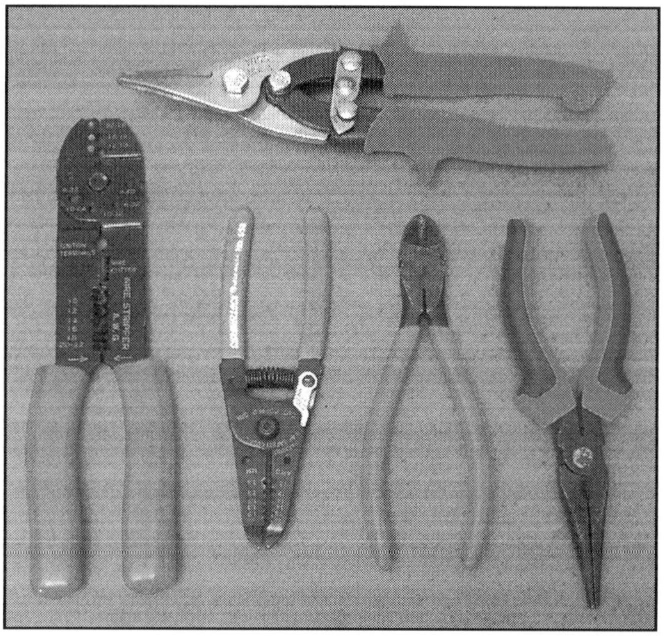

Tools for working with wire

Wire conductors vary in diameter and type of insulation. The core of each wire can either be stranded (many fine wires) or solid (one piece). Stranded wires are more flexible than solid wires and can be easier to work with for certain applications.

Ampacity

The larger the diameter of the wire, the more current (amps) it can carry. The wires used throughout a system must match the intended current flow. The amount of current that a conductor can safely carry is termed its ampacity. The ampacity rating of a wire is based on a number of factors:

- The diameter of the wire
- The type of insulation that covers the wire
- The temperature of its surroundings (ambient temperature)
- How close the wire is to other conductors that also generate heat during conduction of current
- The nature of the raceway or run that encloses the wire, which may allow heat buildup.

Voltage drop and wire diameter

Diameter of wire is also a factor in voltage drop. The longer the run of wire, the more resistance is encountered, which causes a loss in voltage. The resistance can be reduced by using a larger diameter wire. Voltage drop is an important consideration for any long wire run, but is of greater significance from the panel to the battery bank. Using larger wire diameters than necessary is a good idea. This is usually tempered, however, by the cost of the wire. Naturally, the more copper there is in a wire the more expensive it is per foot.

At the very least, all wire conductors in your system must be rated to adequately and safely carry the system's amperage, and should also be sized to minimize voltage drop as much as possible.

Your wire needs to be able to conduct adequately and safely in the environment in which it is used. Some cable can be buried underground, and some cannot. Some can be exposed to sunlight and survive for long periods of time, while others degrade in that type of situation. Some wire does well in damp locations while others do not. Some can withstand heat more than others without degrading over time.

All of these conditions are important to consider. Photovoltaic panel wiring, for instance, is subject to high heat as well as other climate conditions, rain, ice, dew, frost. The cable carrying the current from the panels to the DC service box or batteries is usually buried, so it is important to use cable that is suitable for that purpose.

These are the types of conductors used most frequently for PV installations:

- TC Tray cable
- UFB Underground feed, direct burial
- NM Non metal jacketed for dry locations
- NMC Non-metal jacketed, for corrosive or damp locations, not to be used for direct burial
- USE direct burial service entrance
- Zip cord
- Battery cable
- Inverter cable

Tray cable can be purchased that is both resistant to UV (ultra-violet rays), and can be directly buried. This is used most frequently to connect one panel to another and panels to combiner box.

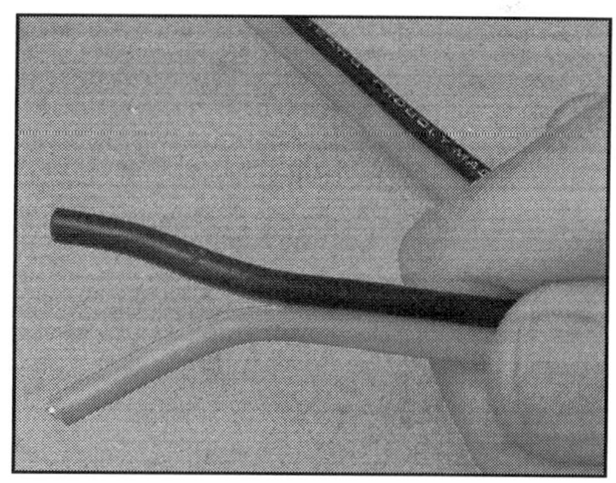

Zip cord

Zip cord is not rated as UV resistant but is often used for panel to panel, and panel to combiner box connections. It is used most frequently to make connections from a DC service panel to batteries, or from a charger to batteries. It is also used for interconnections in DC service panels and other general connections. It is handy to use, flexible and comes color coded as red and black wire.

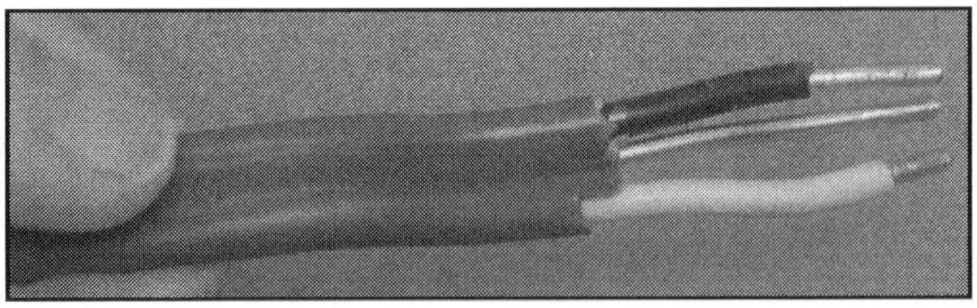

UFB cable

UFB is most frequently used for feed lines from panel combiner boxes to DC service panels and charge controllers. It is UV resistant and can be directly buried.

NM is standard house wiring and used for indoor runs in areas that are not prone to dampness. This wire can be run from an inverter to an AC circuit breaker and distribution box.

NMC can be used in environments that are damp and or corrosive.

USE can be used for feed lines from panel combiner boxes to DC service boxes. It can be directly buried.

Battery cable is used to connect batteries to each other.

Inverter cable is used to connect a battery bank to an inverter.

Battery and inverter cable, sometimes referred to as welding cable, needs to be a large diameter in order to handle higher current. The diameter of inverter cable is based upon how many appliances will be used at any one time. This estimate should include motor startups, which draw large amounts of current for short periods of time. The run or length of the inverter cable should be no more than ten feet between the inverter and battery bank.

Light duty battery interconnects are usually AWG size #4 or #6. These are good for charge and discharge currents of 50 amps or less.

Heavier duty battery interconnects are usually #2, #2/0, or #4/0. As a matter of course I usually use at least #2 battery cable.

Battery to inverter or DC load center are usually #2, #2/0 or #4/0.

SYSTEM VOLTAGE	INVERTER WATTS	INVERTER CABLE
12, 24 or 48	600 to 1000	#2
12	1000 to 1500	#2/0
24	2000 to 3000	#2/0
48	3600 to 4000	#2/0
12	over 1500	#4/0
24	over 3000	#4/0
48	over 4000	#4/0

Resistance and ampacity for copper wire

AWG GAUGE	OHMS PER FT	AMPACITY
0000	0.000049	312.0
00	0.0000779	220.0
2	0.0001563	131.0
6	0.0003951	65.2
8	0.0006282	46.1
10	0.0009989	32.5
12	0.001588	23.0
14	0.002525	16.2
18	0.006385	8.1

Voltage drop

A voltage drop occurs when current flows through any conductor. This is expressed in the formula V = I x R

To find the voltage drop for any length of wire run, multiply the resistance (ohms per foot) of the particular size conductor times the number of feet in the run; then multiply this times the current (amps) that will run through the conductor. The result is the voltage drop. For example, a photovoltaic panel with an output of 4 amps at 18 volts and a wire run of 50 feet to the DC service box through a #10 conductor, will give a voltage drop of 0.19978 volts. This means that at the terminals of the DC service box the system will be delivering about 17.8 volts.

R x ft. x amps (resistance x length of run in feet x amps) = voltage drop.

Sources for wire

ITEM	SOURCE
Zip cord – red/black, 10 and 12 gauge	The RF Connection, All Electronics, automotive or hardware stores
UFB, NM, NMC, USE	Electrical, building supply, or hardware stores
Tray cable	Electrical or solar equipment suppliers

Wire terminals

For the Solar II project we used crimp-on terminals for stranded wire and single conductor wire connections when appropriate. Non-insulated terminals were used, with heat shrink tubing added to provide the insulation where the wire meets the crimp. This

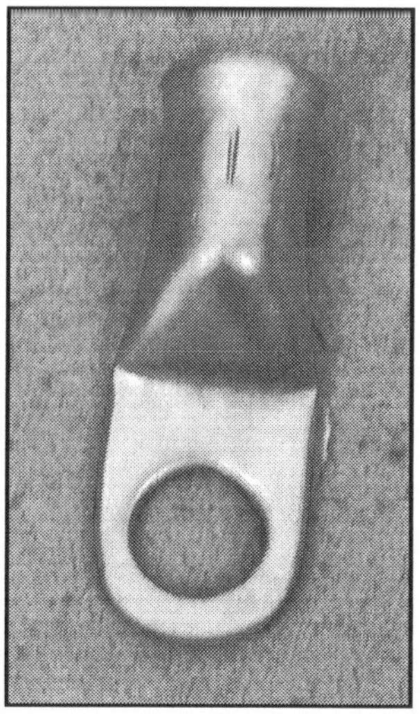

Crimp-on terminal

gives a neater appearance than the pre-insulated terminals. However if you want to use insulated terminals, that is fine.

Crimp-on terminals come in a variety of shapes and sizes. Most of the crimp-on terminals that we used were for terminal block connections and were the spade type for 12 and 10 AWG wire. The sizes you will be using most will be for a #10 stud, #8 stud, and #6 stud.

If you will be adding your own crimp terminals to inverter or battery cable, you will need a heavy duty crimper, and terminals sized for you wire gauge.

Heavy duty crimper for battery and inverter cables

Shrink tube can also be used when splicing wires together. The shrink tube acts as a cover and insulator for the bare conductors. Shrink tube comes in various shapes and sizes and can be purchased at McMaster Carr or any electronics outlet or hardware store. A bag of assorted heat shrink tube is a good investment, so that you can easily choose the best size for any particular connection.

Shrink tube can be made of polyolefin, PVC, or a variety of other materials that have specific use for certain conditions. It also can have adhesive liners that help prevent deterioration of connections from moisture ingress. McMaster Carr has an excellent polyolefin black 45 piece tube assortment.

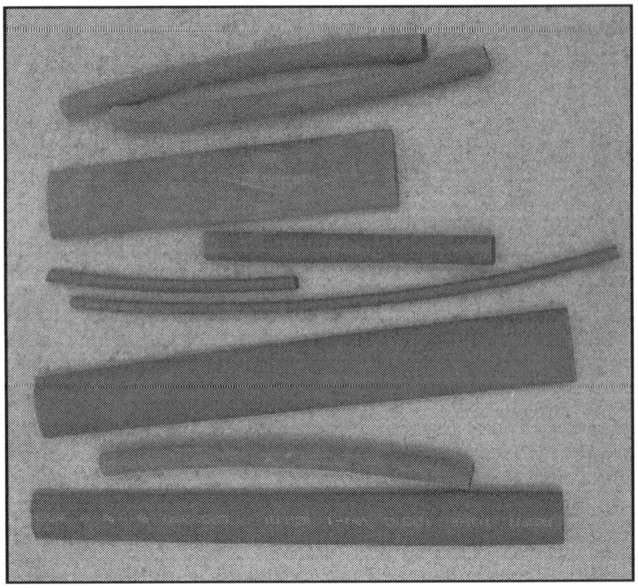

Assorted shrink tubing

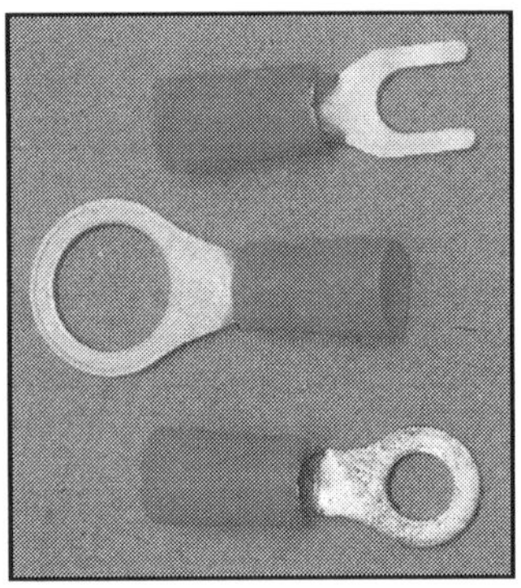

Spade and ring terminals

Batteries have a variety of post terminals. For general connections to these posts, use ring terminals. Ring terminals will not slide off their post connections if the nuts loosen. They come in a variety of widths, thicknesses and hole diameters. Match the hole diameter as closely as possible with your battery terminal screws so that you have a firm, snug fit.

Ring terminals can also be used for any other connection. Most hardware stores, large home centers, electrical suppliers, and automotive stores have ring terminals and spade terminals in stock. If all you can find is the insulated variety you can use them as is, or remove the insulation by twisting it off with a pair of pliers and replacing with heat shrink tubing, if desired.

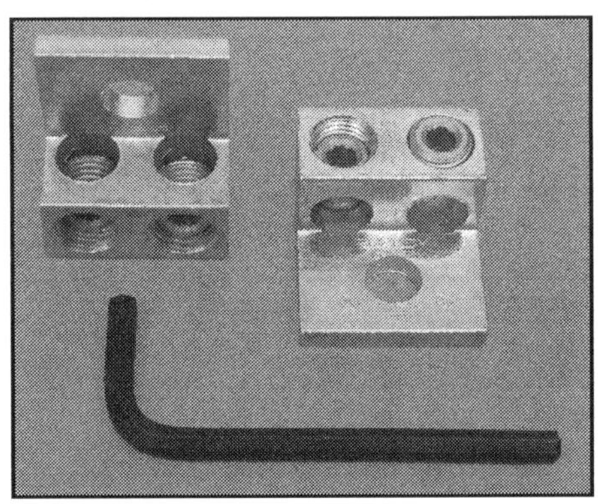

Battery terminal blocks

On some connections you may want to use liquid electrical tape. This is a vinyl liquid that can be brushed right on the crimped connection to cover and insulate it. It is acid resistant and is good to use for insulation around battery banks with sulfuric acid aerosols. It is also good to use if you nick a wire sheath or otherwise have a bare wire or component that you wish to insulate electrically. You can also use a combination of shrink tube and liquid electrical tape.

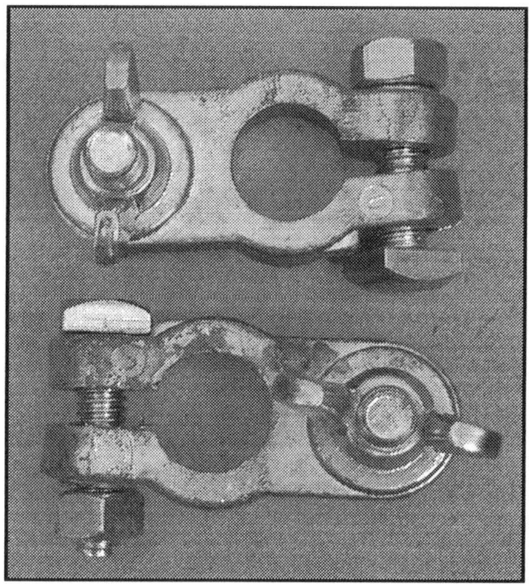

Battery post terminals

Sources for wire terminals

ITEM	SOURCE	ITEM NUMBER
Liquid electrical tape	McMaster Carr, hardware store	#76425A22
Heat shrink tube, assorted lengths or roll depending on needs	McMaster Carr, hardware, electrical, electronics store	#7496K21 (assortment)
Crimp-on spade terminal block connectors, non insulated, 12-10 AWG, #10 stud	McMaster Carr	#69145-K64
Crimp-on spade terminal block connectors, non insulated, 12-10 AWG, #8 stud	McMaster Carr	#69145K63
Crimp-on spade terminal block, non insulated, 12-10 AWG, #6 stud	McMaster Carr	#69145K62
Ring terminals as needed	McMaster Carr, automotive, hardware, electrical, electronics store	

Fuses

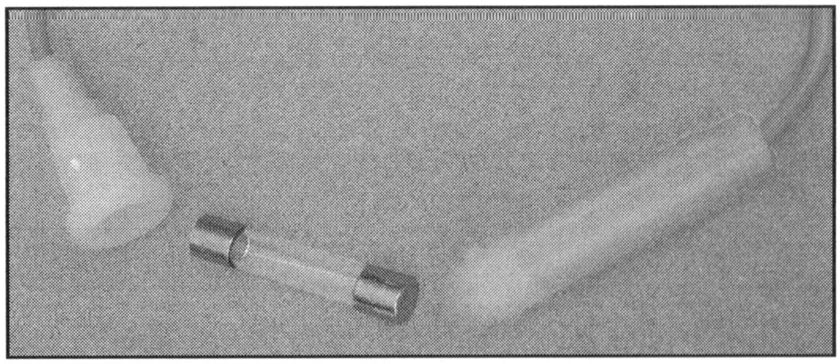

Fuse and inline holder

Fuses provide circuit protection from overload and short circuit current. When an over current occurs, a fusible link within the fuse melts and opens the circuit to prevent conduction.

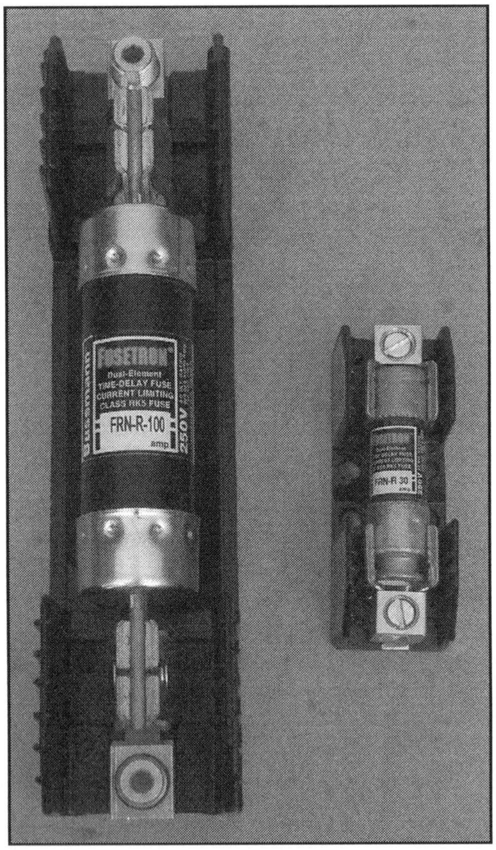

DC fuses

All positive wires from the battery bank to devices need to be fused. Make sure that:

- The voltage rating of the fuses is equal to or greater than the system voltage.
- The ampere rating of a fuse does not exceed the current carrying capacity of the circuit.

A good rule of thumb is to use a fuse rated for 125% to 150% over the amp flow that you expect within the circuit under normal operating conditions; and rated no higher than the ampacity of the smallest wire in your circuit. When considering circuits with motors, take into account the heavy current surge that a motor creates on startup for a brief moment. A higher current rated fuse or time delay fuse can answer to these issues depending on your needs.

Basic Electrical Components

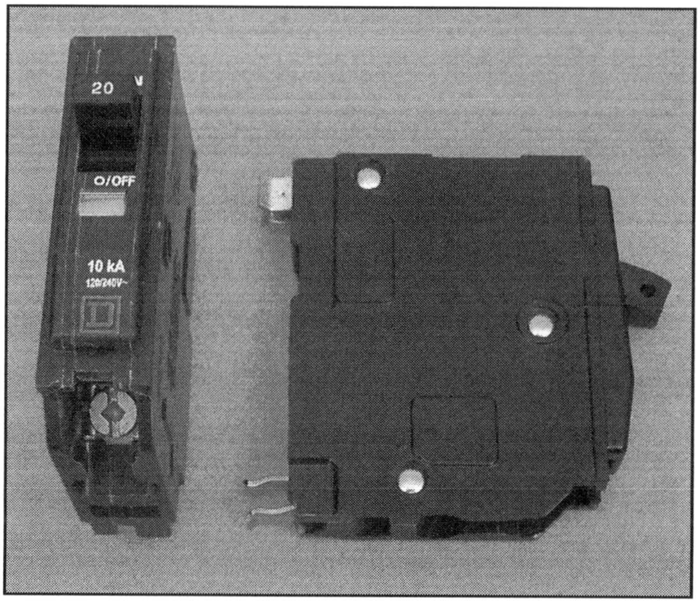

AC circuit breakers

There are many types, shapes and sizes of fuses. Underwriters Laboratories Inc. classifies fuses according to voltage and amperage rating, the degree to which they are current limiting, time delay characteristics, dimensions, and ability to interrupt high fault currents.

Circuit breakers can also be used in DC circuits if rated for those circuits. A circuit breaker is a switch that opens automatically when an overload condition occurs. It performs the same service as a fuse, but can be reset after the problem with the circuit is corrected.

Circuit breakers are used in the AC distribution box for the Solar II system. They are not used in the DC side. They can be used on the DC side but because of tendency of the contacts to arc and fuse under certain conditions with DC, we prefer to not use them.

Diodes, transistors, resistors

There are many types of diodes available for different functions. A basic diode limits current flow to one direction only, and is used in PV systems in either a bypass function or a blocking function.

Diodes used in a blocking function are usually placed between the array and the battery bank to prevent current back-flow from the battery to photovoltaic panels at night when the sun is not shining.

Diodes used for bypass functions are not usually used in 12 volt systems. They are normally used with 24 volt and 48 volt systems where multiple panels are connected in series. Diodes used for bypass are connected to the positive and negative output terminals of each panel. The cathode is connected to the positive terminal and the anode is connected to the negative terminal in the junction box of each panel. If one of the panels in the array becomes shaded or defective, the flow of current from the other panels will bypass that panel so that the defective, or shadowed panel will not interfere with or stop the flow of current to the battery bank.

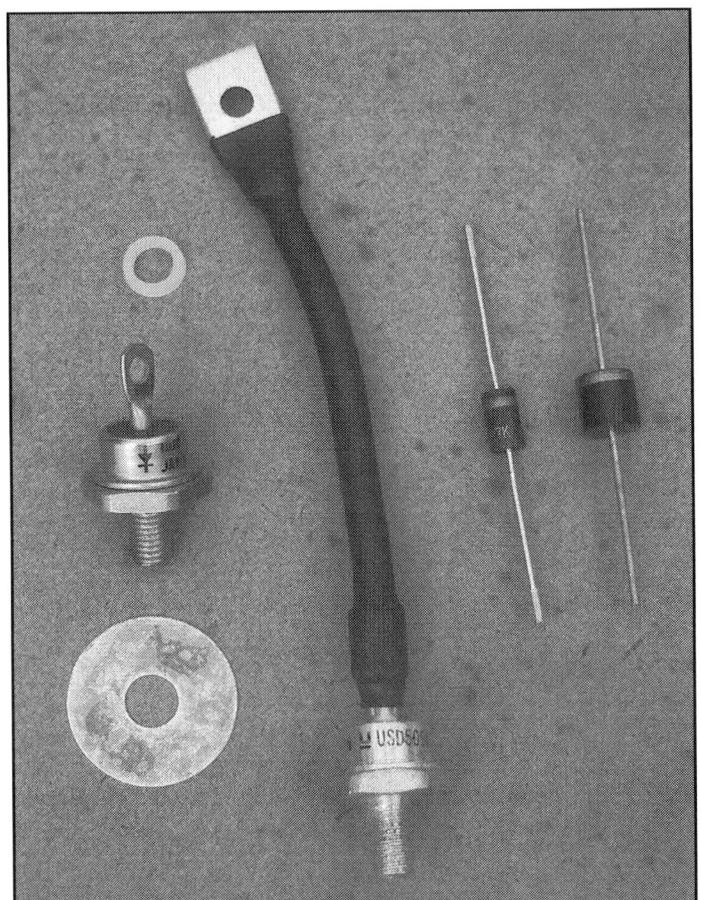

Assorted diodes

Diodes are rated according to PRV/PIV which is the peak reverse/peak inverse voltage. This is the maximum voltage which the diode will block in the reverse biased mode. Practically speaking this means that the diode you use should be rated at above the voltage in the system so that reverse flow does not occur. Diodes are also rated by the maximum current they can carry in the forward biased direction. This means that any diode used in a system should be rated above the maximum current expected for the system.

Diodes come in a variety of shapes and sizes, but the most common are axial lead, and stud mount. Stud mount diodes are contained in a metal case with a screw terminal that can be affixed to a heat sink. Stud mounts are high power diodes that can handle higher currents. They are used with a heat sink as a blocking diode in PV systems.

It is preferable to use Schottky diodes in PV systems. They have less voltage drop than regular diodes.

Other frequently used diodes are the zener and the LED (light emitting diode). A zener is designed to break down and conduct in reverse bias at a specified voltage and is useful for voltage regulation. The LED is designed to emit light and is used in most electronic equipment for visual indicators.

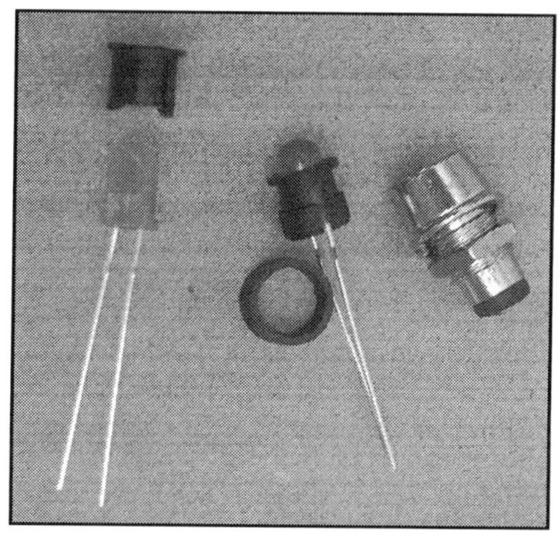

LEDs and holders

Transistors

Transistors are semiconductors with three leads. A small current or voltage at one lead will allow a much larger current to flow through the other two leads. Transistors are usually used as amplifiers and switches. There are two basic types: bipolar and field effect.

Power transistors such as the Darlington used in the CC2 controller for the Solar II project are frequently used in voltage regulators as switches. Power transistors generate a lot of heat in use, so the metal case of the transistor needs to be connected to a very large heat sink.

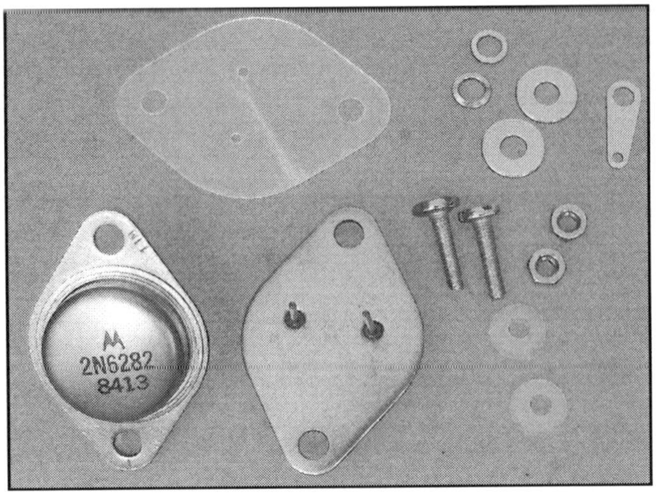

Transistor kit

Resistors

Resistors reduce current flow and have many applications in electrical circuits such as current limiters, voltage dividers, and current dividers. They can have a fixed resistance or be variable. Thermistors are temperature sensitive resistors that are commonly used in charge controllers to adjust charging to the batteries based on the temperature of the battery. Photo resistors are commonly used in lighting controls and other functions.

For in-depth information about diodes, transistors, resistors and other electrical and electronic components, see *Lessons In Electric Circuits* web page by Tony R. Kuphaldt, or purchase some of Forrest Mims easy-to-understand texts on basic electricity and electronics. Links can be found in the resources section.

Heat sinks

Diode and power transistor heat sinks should be sized according to the maximum expected watts generated by the PV array. Heat sinks are rated in watts for their heat dissipating capacity. If a system has a short circuit current of 3.5 amps and closed circuit voltage of 18 volts, the designer watt rating for the array with four panels connected in parallel would be 252 watts. This means that it needs a heat sink rated at 252 watts or more. A 300 watt heat sink would be a good choice as it is always advisable to use a larger heat sink than needed.

Designer watts are based on short circuit current measurements and closed circuit voltage measurements under 1 sun conditions. Designer watts do not equate with real time conditions but are a guideline for selecting components. Some codes suggest an additional 30% be added to the designer watts for heat sinks. Although this is not necessary, it is a good practice. So, if a system is putting out 252 designer watts, I would add 30% and provide a heat sink that is rated at about 327 watts.

Switches

Switches come in many different styles for different circumstances. Switches can be controlled manually, such as a toggle, rocker, push button, or knife type switches. They can also be controlled by light, magnetics, voltage, current, temperature and other means.

Each switch style has variety of configurations. The most common are single pole single throw, single pole double throw, double pole single throw, double pole double throw. Each of

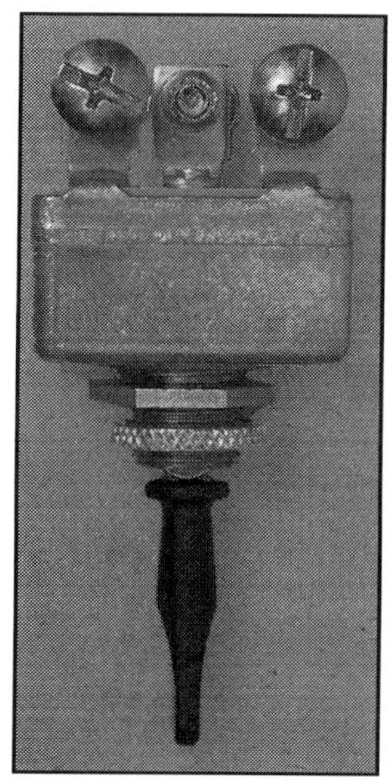

50 amp DC 12 volt single pole double throw toggle switch

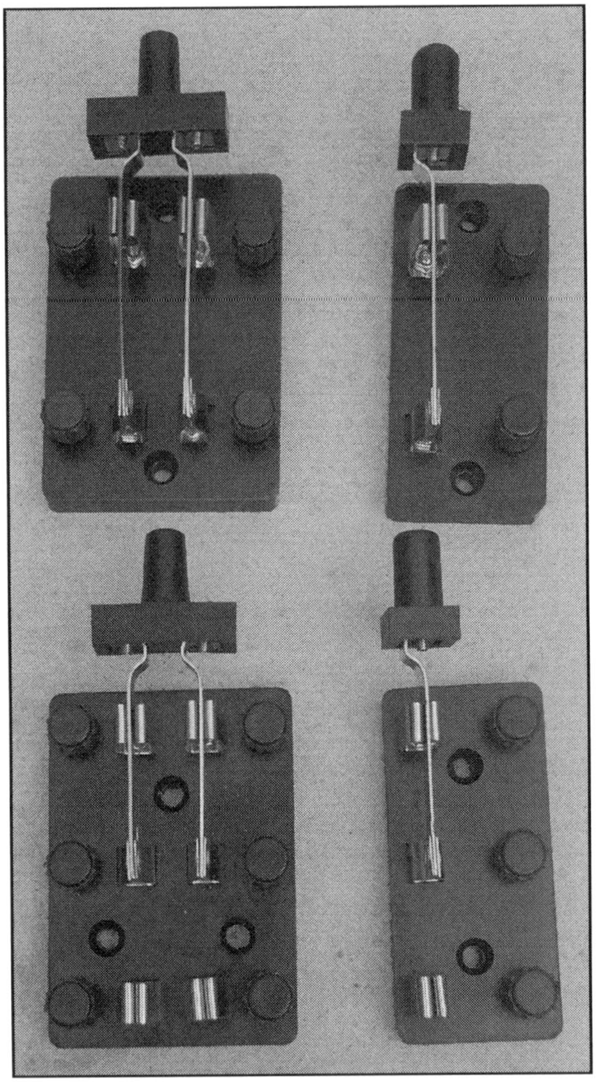

From top, left to right: double pole, single throw single pole, single throw double pole, double throw single pole, double throw

these throws provides different capabilities for the switch. In most switches it is hard to see the action of the switch represented, but some examples of these different throws in the knife switch style will give you a good idea of the possibilities such as in the photo at left

Switches are rated for either AC or DC, or both. They are also rated for current carrying capacity and voltage. A switch must be able to carry the current and voltage the circuit requires. It is standard practice to use switches that are rated for higher currents and voltage than needed in a circuit, to add a margin of safety. It's worth it to purchase the best switches available and pay a little more, rather than to opt for low cost switches. Switches are important components and the risk of having them fail is not worth a few pennies saved up front.

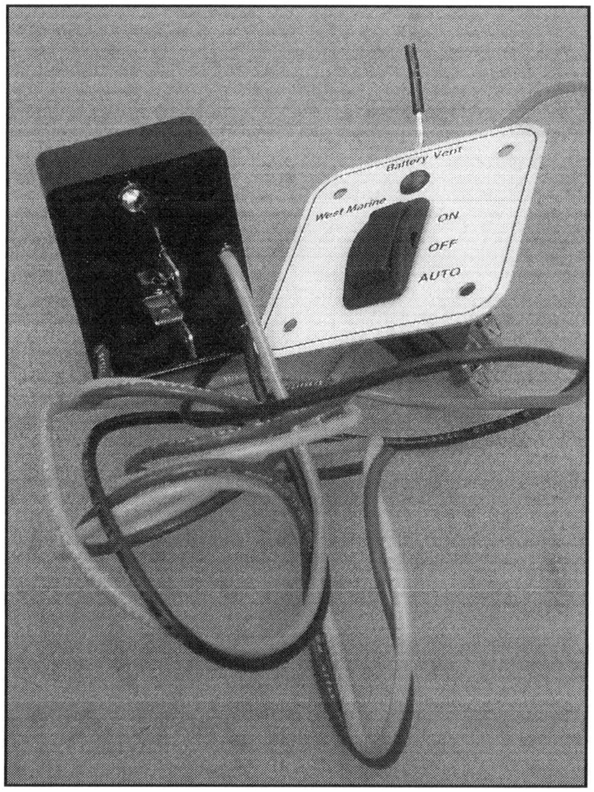

Battery vent rocker switch

Basics of Mounting Solar Panels

Where and how to mount panels is of prime importance. Panels should be:

1. As close as possible to the location of power use,
2. Free from all shade,
3. As accessible as possible for easy maintenance,
4. On racks and mounts that are very securely attached and strong,
5. On frames that are grounded electrically.

1. Panels should be as close as possible to the location of power use.

Long runs of wire create a drop in voltage. Minimize voltage drop by positioning the panels as close to the point of use of power as possible. Short runs also minimize the possibility of induced surges from nearby lightning strikes. Another benefit is that you can use the minimal wire diameter necessary. With a longer run, larger diameter (and thus more expensive) wire must be used to mitigate voltage drop.

Another way to deal with long wire runs from the array to the point of power use is to wire the panels in a 24 volt or 48 volt configuration. This will permit a longer run without the excessive voltage drop. The batteries would have to be configured to match the panel array, and if you have an inverter it would also have to match the rest of the system voltage. If you are using DC power without an inverter, you would need to add a voltage converter to step down the voltage so that you could run 12 volt appliances on the system. Please note that a 48 volt system is in the serious shock hazard range. I suggest staying in the 12 volt range until you have some experience putting these systems together and operating them.

2. Panels should be free from all shade.

Absolutely avoid siting panels in any location that has full shade during any part of the day. If there is any shade on your panels, you will have to install bypass diodes on each panel. If you have a choice, go for the most shade free location.

3. Panels should be as accessible as possible for easy maintenance.

You would be surprised at the amount of dust and pollen that can accumulate on PV panel surfaces. In some locations, this accumulation requires at least a weekly cleaning with some water, especially if there is little or no rain. That very fine layer of pollen and dust will reduce panel output if left to accumulate.

It is also helpful to be able to easily view and access wires and connections to monitor and fix any problems that may need to be attended to. If the system is very high up on a pole or on a roof, such maintenance is not so easy.

On the other hand, there may be overriding reasons for mounting the panels high and out of reach, for instance, see the photo on page 65.

4. Panel racks should be very securely mounted

Secure panel foundations are a must. Solar panels behave like sails in high winds, and strong gusts can dislodge them and frames and easily destroy a system. Anything ground mounted needs to be well anchored to the ground.

5. Panels and frames should be grounded electrically

Assume that your panels and frame are a lightning magnet, and treat them accordingly by including grounding for the frame and panels. Even non-direct nearby lightning strikes can induce heavy currents in a conductive structure, so make sure that lightning can be safely discharged to ground.

Options for mounting solar panel arrays

Racks for stationary PV arrays are either mounted on a building, or are free standing.

Roof mounts

Many people like the idea of roof mounting PV panels because it uses an already existing structure; and/or it gets the panels up and out of the way, if yard space is limited. Roof mounted panels are generally on racks that are bolted through the roofing, or mounted on a pole that goes through the roof and is anchored in the building.

When mounted directly on the roof surface, there is usually a pair of legs on the upper mounts to give the panel the correct tilt. In cooler climates, snow and ice build up can cause severe problems with roof surface mounted panels if the pitch of the roof is not steep enough.

photos courtesy Independent Power & Light

Roof mounted panels and detail

If you're interested in putting a PV array on your roof, consider the options very carefully. Installing roof mounts requires a certain amount of carpentry expertise so that the roof (and the entire building, in the case of pole roof mounts) is not compromised. As many homeowners know, putting a hole in a roof can be a nightmare if you don't do it well.

Roof mounts are fine when they are installed correctly. Remember that if problems occur, you will need to get up on that roof and fix them, usually in the worst weather.

Pole roof mount

photos courtesy Independent Power & Light

Wall mounted panels, summer position and winter position

Wall mounts

If you have good snow cover in the winter and have a building with a south facing wall, another option is wall mounting the array, as shown above. Snow cover is reflective enough so that the panels can be vertical without reducing performance in the winter. This also eliminates the need for regular snow removal from the panels.

photo courtesy Independent Power & Light

Top of pole mount

Free standing mounts

As you may have noticed by now, I have a personal bias for low, ground mounted frames. Even though I have a forty foot tower next to my house and a very large aluminum flagpole lying around, I have never tried to mount any panels on them.

Tall pole and tower mounts are called for when there are obstacles you need to rise above, such as trees you do not want

photo courtesy
Independent Power & Light

Moose proof mount for PV array

to cut; or when you need to make your array less accessible to ruffians in the neighborhood, such as the moose proof pole mounted array in the photo at left. Pole mounts are either top of pole, side of pole, or tracking mounts.

Low four point ground mount frames are easy to build with locally available materials, and they're usually simpler to anchor to the ground than pole mounted (single point) frames, especially when there is ledge near the soil surface. Tall pole and tower mounts require very substantial foundations and deep holes.

Keeping the frame low keeps your solar array readily accessible for periodic inspection, maintenance, and cleaning, and I personally find that if it is not very convenient (if I have to climb up to a high roof, pole or tower) I am not likely not keep up a good routine of inspection and maintenance. However, if you live in a region with a fair amount of snowfall, be sure that the bottom of your panels is above your highest snowfall accumulation.

photo courtesy Independent Power & Light

Four point ground mount

Building a rack for the PV array

When bad things happen to good panels

Panel racks need to be weather proof for your climate. Rain, wind, snow, heat and cold take their toll over a period of time, and if you do not use the correct materials for your environment, you will sooner or later have a rack failure. This can be catastrophic if you have many thousands of dollars worth of panels on such a rack. You can expect your panels to have a lifetime of 20 years or more, so either the racks should be designed to last that long, or you should plan to build replacement racks and remount as needed, definitely before the panels come crashing down. Moisture in the form of dew, rain, melting snow and ice, can cause corrosion on metal, and deterioration via expansion and contraction in porous materials such as wood, not to mention rot in wood. Extreme temperatures cause brittleness and cracking in some materials, as well as warping and general deterioration of materials as they are exposed to the constant daily expansion and contraction cycle of heat and cold throughout the seasons.

The original racks failed, but these old panels are still working.
photo courtesy Independent Power & Light

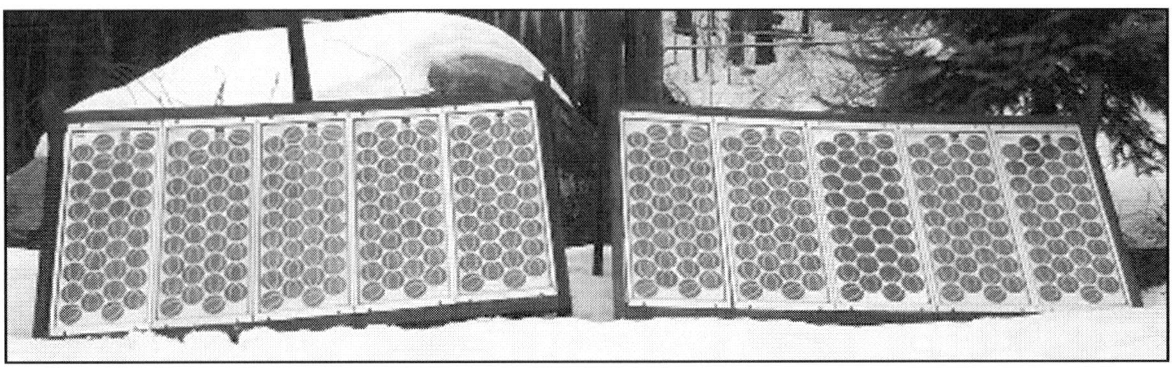

The aluminum stock in this test rack was not strong enough for gusting winds.

Wind stress can easily buckle and twist frames and stands. Wind damage is probably the single greatest cause of rack failure and panel loss. PV panels act as sails and in high winds can produce a powerful lift. The force involved should not be underestimated.

The portable test rack in the photo is an example of a rack that had some aluminum stock less than ⅛" thick. Five minutes after I completed this rack and attached some panels, a raging, but brief 10 minute storm came by with about 40 mile per hour winds, and completely bent the hypotenuse of my rack, which told me never to do that again. Although I have bent it back into shape and still use it occasionally for testing panels, I am cautious if there is any chance of high wind.

Rack materials

The best material to use for the structural members of the panel frames is stainless steel. Second best material is galvanized steel, and third choice is aluminum. Next is pressure treated wood, followed by wood that is painted with a sealant such as acrylic. Your choices of materials will depend on your budget. Despite budget constraints, some options should only be implemented with caution or reserve.

Wood

Wood, for instance, appears to be the least expensive rack material, however, it deteriorates quickly from moisture, and unless well sealed will not last very long. The proper wood coatings are expensive. You could use pressure treated lumber, however this is more expensive than regular mill lumber.

Regular mill lumber must be re-coated regularly to keep the moisture out. Any coating you use should be UV resistant, such as most outdoor latex paints or acrylic coatings. A good UV resistant coating is more expensive but it is a must for the harsh environment encountered by panel racks.

If you do build wood racks, use the best fasteners that you can afford. Best is stainless steel nuts, screws and washers. The next best is galvanized steel or 18-8 steel. If you do use fasteners other than stainless steel, coat them with silicone caulk. I do not recommend using nails in a rack. This does not mean that a good carpenter couldn't a build good rack with wood and nails, but the use of screws will allow easier disassembly if desired at a later date. Many times a design flaw is not

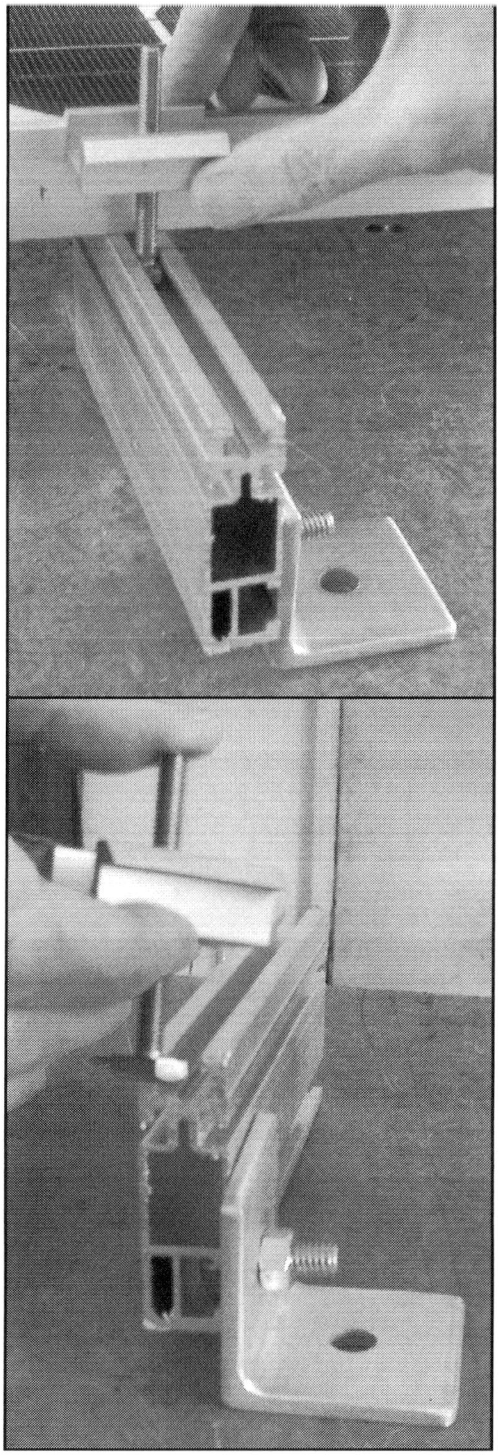

Some components of a Unirac kit for roof mounting.

immediately apparent, and if your are experimenting with rack designs you'll be glad to have fasteners that can be easily undone.

Galvanized steel

Galvanized steel angle stock is readily available at the local hardware store. I consider it to be one of the best options for building racks as it is easily available and is not that costly considering the longevity and strength of the materials. Slotted steel angles are convenient to work with as there are ready holes with which to fasten your pieces together. This saves a lot of work drilling holes except for the occasional widening of a hole when the dimensions don't exactly match. Non-galvanized steel can be coated with a rust preventative primer, but this will require regular re-application of the primer as needed.

Aluminum

Aluminum is easy to cut and drill, and is lightweight. If you do use aluminum however you need a thick stock.

I personally would not use anything under a ⅛" thick stock on any panel frame component.

If you have metal working skills and can weld, you can avoid the use and cost of fasteners, and make a more tailored frame. However for the rest of us, I recommend using galvanized slotted steel angle or thick aluminum stock that is easy to work.

Another option is to have a metal shop fabricate a rack for you. The added expense may be worth it, if you do not wish to take the time to build a rack. And of course, commercial rack kits that are specifically made for PV panels can be purchased, but keep in mind that if you have made the panels yourself, they may be heavier than commercial panels. Be sure that any racks you purchase can carry any extra weight. The commercial rack kits can be expensive, but they're really easy to put together, and if you have purchased PV panels, there is probably a kit for your specific panels.

Solar panel rack foundations

For a permanent installation, some type of concrete foundation is advisable. The purpose of the foundation is to provide a level, solid footing and support for the panel rack, as well as weight so that the PV array does not blow away in high winds.

You can use 6" diameter cardboard construction forms to make a pier foundation, you can make a concrete slab foundation, or you can use concrete construction blocks or other types of pre formed concrete block. Slab foundations need to be thick enough not to crack due to contraction and expansion from extreme temperature fluctuations. Pier foundations should be it at least 2' deep. Although it is recommended for piers to be at least as deep as the frost line, I have found that two feet is sufficient in most cases.

If you use concrete construction blocks, the depth will be limited to the length of the block which is usually about a little over fifteen inches. Some of the block will be above ground so the depth will be a little less than fifteen inches, which is sufficient in most situations. You could also bolt two concrete blocks together to get more depth and weight.

If you pour a pier or slab foundation you will need to insert, position, and level L anchor bolts, or threaded rod, to align with the attachment holes in the rack feet. Construction blocks or other type of preformed concrete blocks will require a masonry drill to drill holes for lag screws, anchor bolts or attachment screws.

Whatever type of foundation you make, it will need to be level, weighty and of course, positioned so that your PV array is correctly oriented.

A four point ground frame panel rack

For the Solar II project we decided to build a four point ground frame panel rack. The needed materials were available at the local hardware stores, the level of expertise needed to construct it was minimal, and the result is a rack that is sturdy and economical.

This rack was designed to hold the array of four 29" x 36" panels featured in the instructions in *Build Your Own Solar Panel*. The rack materials used were cement blocks, galvanized steel slotted angle ⅛" thick by 1½", and an assortment of bolts, screws, and washers.

Steel angle is readily available at most hardware and building supply stores. Sometimes the exact lengths you want are not in stock when you need them. This is not a problem. You can bolt several pieces of angle together to make the length needed, and cut it to length with a hacksaw if necessary. I needed two 8' lengths but I couldn't get 8' lengths locally, so I purchased 6' and 3' pieces and overlapped them 1' to get the 8' needed. If you need to piece lengths of angle like this, overlap and bolt them first, then if necessary, cut them to the exact length you need. Of course the best option is to purchase the exact length so that you do not have to fabricate extensions.

The rack is a simple assembly that can be set at any angle desired. The 8' hypotenuse rails (the part of the rack that is moved to adjust the angle of the array) and the horizontal base rails are connected by a 5⁄16" bolt that acts as a swivel pin. It is a simple, relatively low cost, and most important, sturdy rack.

For the horizontal base of the rack, we used a pair of 6' pieces of angle. This angle was bolted to four concrete blocks purchased at the local hardware store.

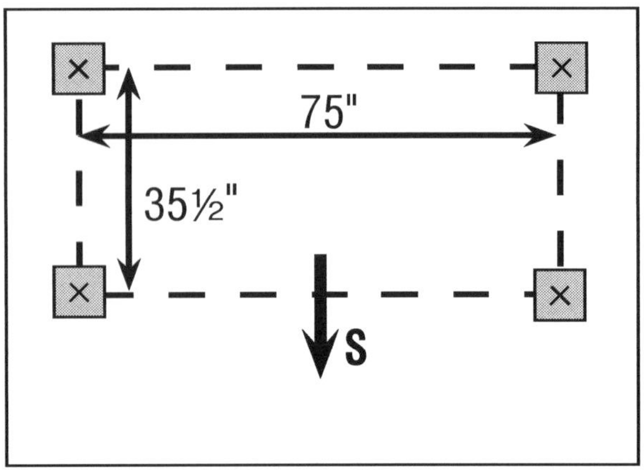

Placement for the concrete foundation blocks

Preparing the site

To prepare to assemble the rack, we had to orient the site, dig the four holes for the blocks according to the orientation, and place and level the blocks in the holes. The holes for the blocks were dug large enough so that

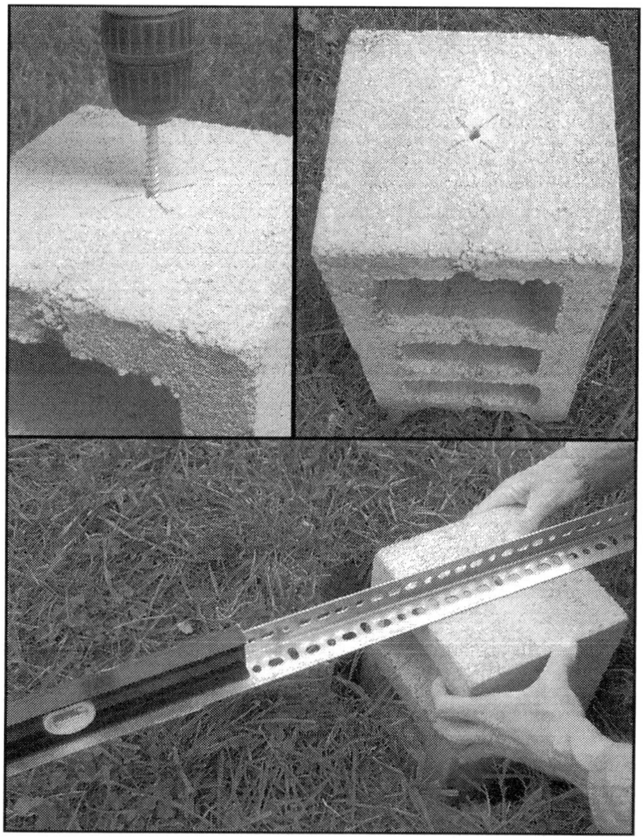

Top, drilling a hole in the center of the block; bottom, adjusting the block alignment

the blocks could be adjusted to get proper alignment.

Each base rail is attached to two blocks. The distance between the two base rails must accommodate the solar panels that you are going to mount. Four 6' cross rails would be used to mount the panels, so the attachment holes in the blocks for the two base rails needed to be 75" apart (east-west distance), measured from the centers of the concrete blocks. The north-south distance between the blocks is 35½". After the holes were dug, a hole was drilled with a concrete bit to accommodate a 5⁄16" bolt in the center of each concrete block. The blocks were then placed in the holes, the angles were laid on the blocks, and the holes in the blocks were aligned with the holes in the angle. The blocks and holes were adjusted to make everything square and level, and correctly oriented.

Assembling the rack

The two 6' pieces of angle for the base rails were attached to the blocks with 5⁄16" bolts nuts and 1½" diameter washers, with washers

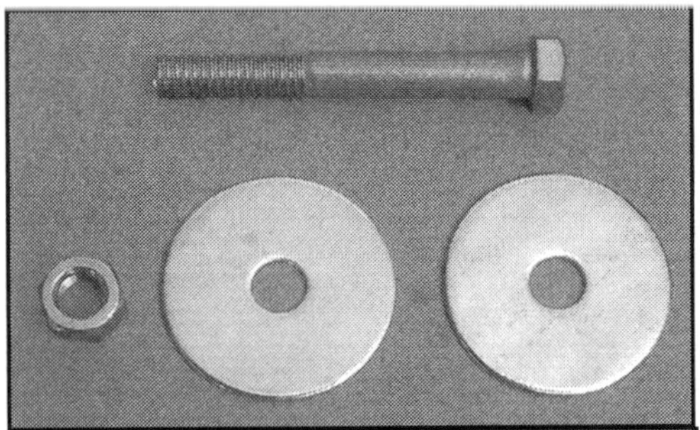

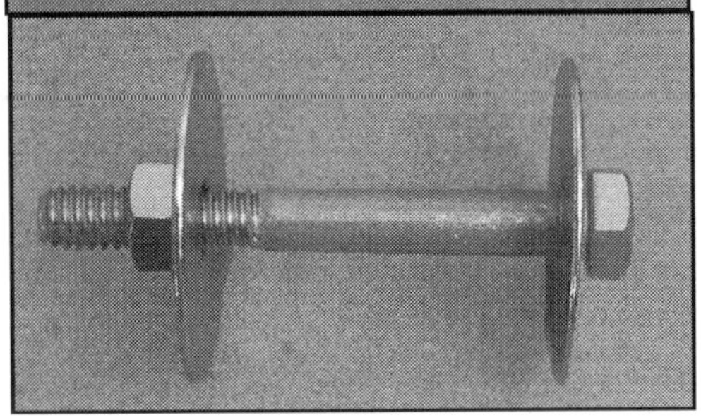

on both sides of the concrete block. The washer on the inside of the block rests against the concrete, and the washer on the outside is on top of the angle. Another washer can be added between the block and the angle, if desired.

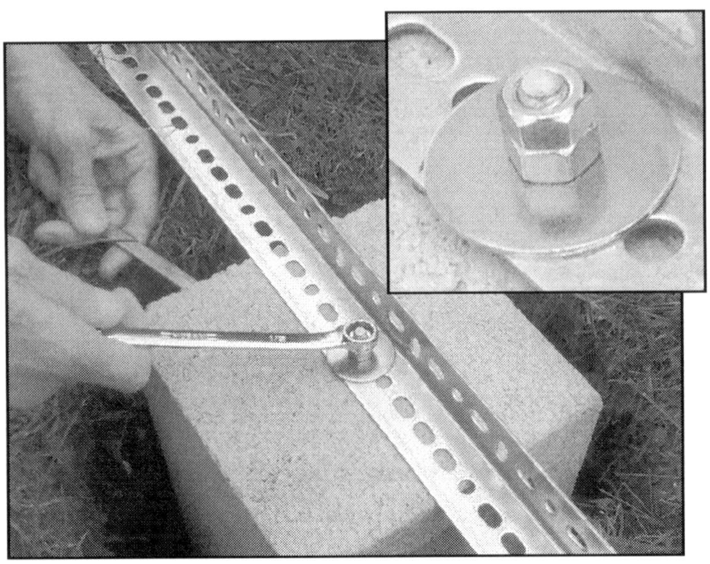

Two nuts are used on top. The extra nut is a locking nut that will hold everything tightly in place. These nuts are tightened as securely as possible, but without so much pressure on the concrete block that it might crack. Hex head screw bolts do not have a fully threaded body, so the threaded length needs to be the right size to tighten everything securely. You can also use pan head fully threaded screws. The point is there should be no loose or moving parts.

This is not the only way the angle can be secured to the block but it's probably the easiest and it works well. If you are using preformed slabs or poured concrete, you will have to use appropriate fasteners, such as L bolts and or lag bolts.

Fastener assembly for attaching rails to rails

Once the two foundation rails are fastened to the blocks, the two hypotenuses, which each consisted of a 6' and a 3' piece of galvanized steel angle bolted together to be 8' in length, were then attached to the base rails. These are the sides of the frame that holds the panels. The two angle sections were bolted together with ¼" or smaller pan head screws and washers. Each hypotenuse

Checking the angle of the rack. This is the summer position, in this case, 30°.

rail was connected to the base rail with a 5⁄16" bolt and several washers. These allow the rack to be raised or lowered to any angle desired for seasonal changes.

Once each hypotenuse rail was attached to its base rail, the vertical rails were installed.

Two pairs of steel angles, one 36" long, and one pair 72" steel angles are used for the vertical rails so the pairs can be exchanged for the different seasonal panel orientations needed. By attaching the vertical rails at different points, the rack can be adjusted to match the seasons and catch the most light. You can find the correct angle by using a protractor and a plumb bob, then mark the slots on the base and hypotenuse rails for future reference.

The pair of vertical rails that were going to be used

immediately were attached to the to the base and hypotenuse rails with screws, washers, and nuts.

Four 6' pieces of the galvanized steel angle were installed as horizontal rails connecting the two hypotenuse rails. These are the rails that the panels are directly attached to.

The base rails were checked again for level and alignment, then the holes and spaces in and around the blocks are filled with dirt in this case, though we could have used mixed concrete.

The vertical rails are connected to each other by a back brace, which consists of a 6.5' wooden 2x4. This back brace adds stability to the whole structure. Four holes were drilled in the 2x4 to accept 5/16" bolts, which attach to the vertical rails. Later, the 2x4 was stained and sealed with several layers of plastic coating to weatherproof it.

Once the back brace was installed, the panels were attached to the rack.

The panels constructed from the instructions in the *Build Your Own Solar Panel* book have

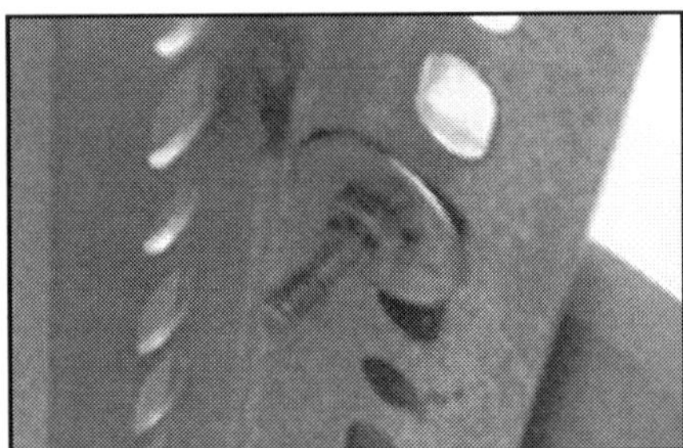

long screws built in to them, so it was easy to mount them in the holes in the angle stock. It was necessary to widen some of the holes in the angle so that the screws and holes would align, but that was fairly easy to do with a mill end and a hand drill.

This is just one way to build a rack. The great thing about slotted angle is that you can play with it like an erector set and come up all kinds of design variations to match your needs and panel sizes.

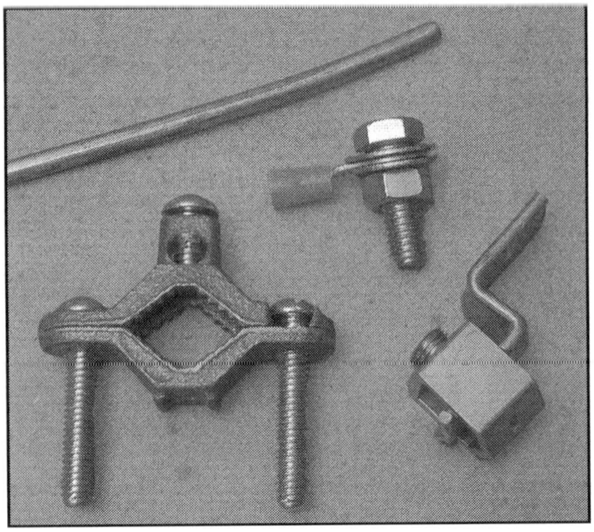

Grounding the panel rack

The panel rack can be grounded from any point on the steel angle rails. The parts used for this project were a ground rod, ground clamp, solderless connector, 6 gauge wire, and some nuts, bolts, and washers.

If you cannot drive a ground rod very deep (8-10') into the earth, there are alternative grounding methods. In any case you will have to dig down to the ledge over a wide area and spread out a series of wires in a radial configuration, or use copper screen or copper sheet.

Ground clamp is attached to the ground rod.

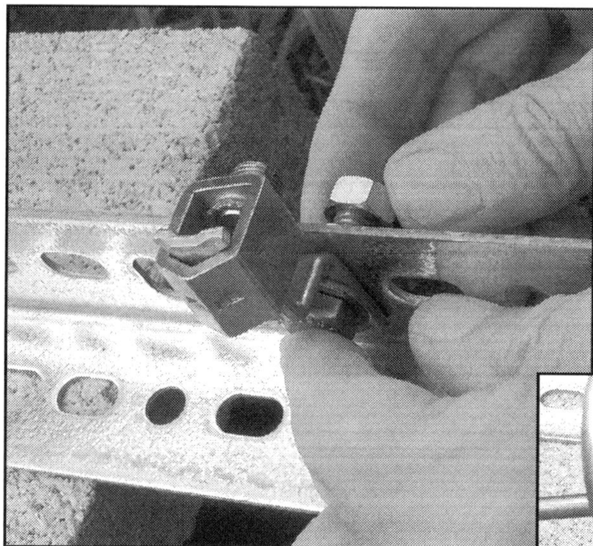

Solderless connector is attached to a base rail, and the 6 gauge wire is bent and cut to size to connect the ground clamp to the solderless connector.

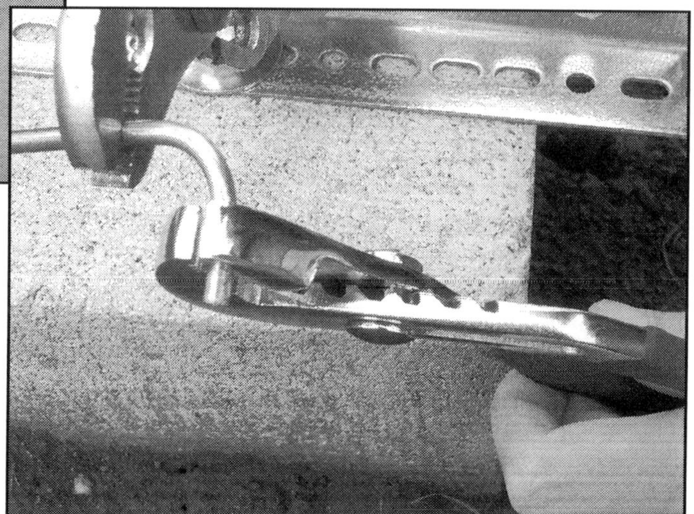

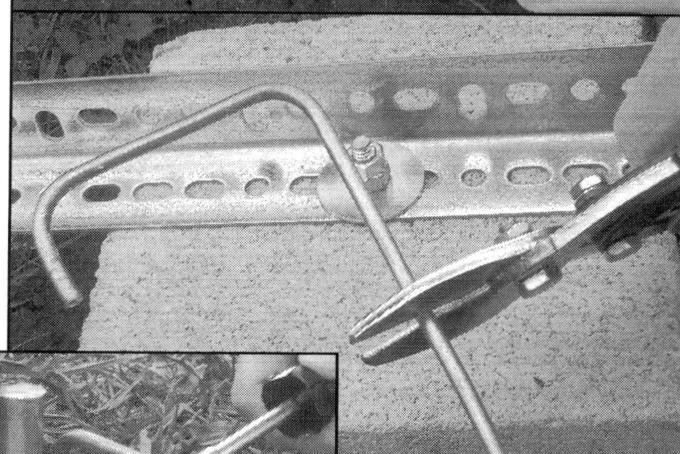

Whatever your grounding method, make sure that all parts are well connected and bolts tightened adequately. The purpose of grounding is to create a good conductive path from all metal parts on the panels and rack to the ground, which will help to avoid damage from either direct lightning strikes or induced current from nearby lightning discharges.

Tools used for rack construction:

Electric drill
Masonry drill bit
(to accommodate 5⁄16" bolt)
Hacksaw
Pliers
Screwdriver
Large hammer to
drive ground rod
Shovel

Following are the parts used to construct this particular rack, including 2 pair of vertical rails for changing the panel angle seasonally. All parts were purchased from the local hardware store.

ITEMS	HOW MANY
Galvanized slotted steel angle, 8" X 1½", 6 ft. (2 base rails, 4 horizontal rails, 2 vertical rails for winter position)	8
Galvanized slotted steel angle, 8" X 1½", 8 ft. (2 hypotenuse rails)	2
Galvanized slotted steel angle, 8" X 1½", 3 ft. (2 vertical rails)	2
Stainless steel washers, 1½"	32
Stainless steel hex head bolts 5⁄16", (to attach base rails to concrete blocks, and cross bars to vertical rails)	8
Stainless steel hex head bolts 5⁄16", (to attach rails to rails)	14
Various bolts, screws nuts, washers	as needed
Lumber - 2x4s, 6.5 ft. long	2
Cement blocks	4
Ground clamp, solderless connector, connecting screws, nuts, washers	as needed
Ground rod	1
Ground wire #6 AWG	as needed

Combiner Box

photo courtesy Independent Power & Light

Combiner box for a pole mounted array

The combiner box can be situated anywhere, but the most efficient placement is on the rack that will be holding the PV panels.

If the rack is made of slotted steel angle, there are plenty of choices for installing the combiner box. The best placement is on the rack behind the panels so that the panels protect the box somewhat from the weather. It's not necessary, but it's good array design to protect and cover wire connections as much as possible.

Combiner boxes must perform well in harsh outdoor conditions, and only certain types of boxes are suitable for this. The combiner box enclosure should keep out dust, rain, ice, snow, and heavy moisture, all of which will deteriorate the service of a PV system.

Combiner box mounted on slotted steel rack

Electrical enclosure specifications

The National Electrical Manufacturers Association (NEMA) designates enclosures for certain types of duty. NEMA type 3 or 4 boxes are weather resistant. Type 3R is rain tight, type 3S is dust and rain tight, type 4 is water tight and dust tight, and type 4X is water tight, dust tight, and corrosion resistant. All 3 and 4 types mentioned will protect from sleet, snow and ice formation.

If dust protection is not needed, a 3R type box will provide protection from the basic elements of weather. If dust is a concern, use a 3S or 4 type box.

You can use anything for a combiner box that will protect the electrical connections from the weather. I have seen some very creative examples of combiner boxes such as short lengths of PVC pipe with screw on caps. Most hardware or building supply stores will probably have some kind of appropriate box in the electrical section.

For the Solar II project we used a 6"x 6"x 4" Cantex PVC junction box, NEMA 3R with rubber gasket purchased at the local hardware store. It was just the right size to comfortably fit the terminal block and lightning arrester. This particular box used four screws to hold the cover.

Option for quick access

This was not an important consideration for us, but you might prefer an enclosure that can be opened and closed quickly and easily, with a hinged door. For instance, you could install a double pole, double throw switch in the combiner box to act as a cut off on the panel output to the battery bank. Some jurisdictions require such a cut off switch at the panel. Such a switch can be installed in a separate box connected to the combiner box, but inside the combiner box is also a good place for it.

If you plan to use the cut off switch a lot, or want to be able to access it quickly in an emergency, a hinged cover will give you swift cut off capabilities. If you do choose a box with a hinged lid, make sure that the box is good quality and that the gasket will seal the interior from the weather.

Planning the combiner box

All the wires from the PV panels converge in the combiner box. The four PV panels in the array are connected by a six position terminal block inside the box. This block has 12 screw terminals and was fitted with four jumpers so that the panels could be connected in parallel.

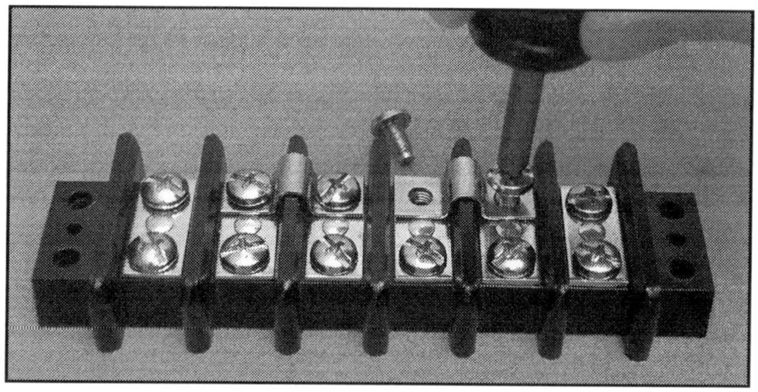

Installing jumpers on the terminal block

Lightning arrestor

A DC lightning arrester for the circuit was included in the box beside the terminal block and connected to it. This particular arrestor comes with a black negative wire, a red positive wire, and a green ground wire. The red wire is connected to the PV positive terminal and the black wire is connected to the PV negative terminal. The green wire is connected to the panel rack with bolts and washers to secure it tightly. The panel rack ground at the foundation rail completes the grounding circuit.

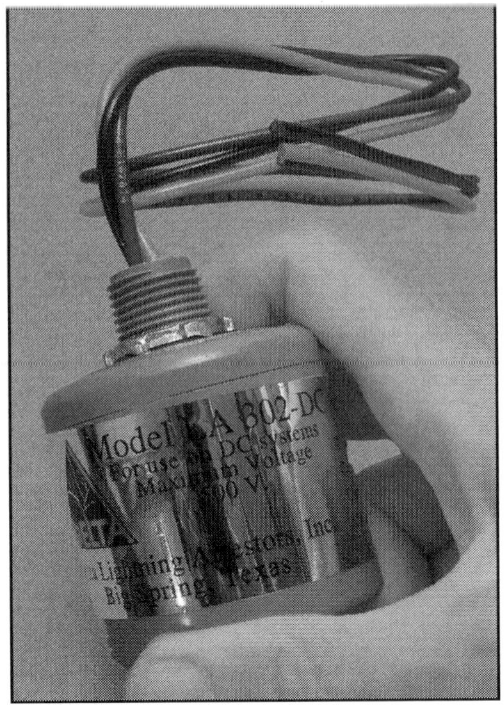

Lightning arrestor

The lightning arrestor used was a silicon oxide varistor lightning arrestor, Delta LA-302DC. The arrestor can be placed in the combiner box, or at any point before the charge controller. You can also put in two arrestors, one at the combiner box and one in the DC service box, or place just one arrestor in the DC service box.

The arrestor does not have to be placed in an enclosure. It comes with a threaded neck, so the PVC canister can be screwed into a tapped hole you provide in the combiner box. It can also be placed elsewhere outside of the combiner box, on or in the vicinity of the rack. Although the arrestor is weatherproof, I prefer to provide them with some kind of enclosure.

Please note that there is no such thing as 100% lightning protection. Although the arrestors are rated to safely conduct many thousands of amps from a strike, if you get a direct hit, it may burst the canister. If this happens, simply replace it with another arrestor.

Assembling the combiner box

Exact placement of the lightning arrestor and terminal block was worked out, so that the best location for the two wire ports could be determined. The wire entry port had to be large enough for four pairs of 12 gauge zipcord wires (one pair for each panel). The output port would have to accommodate

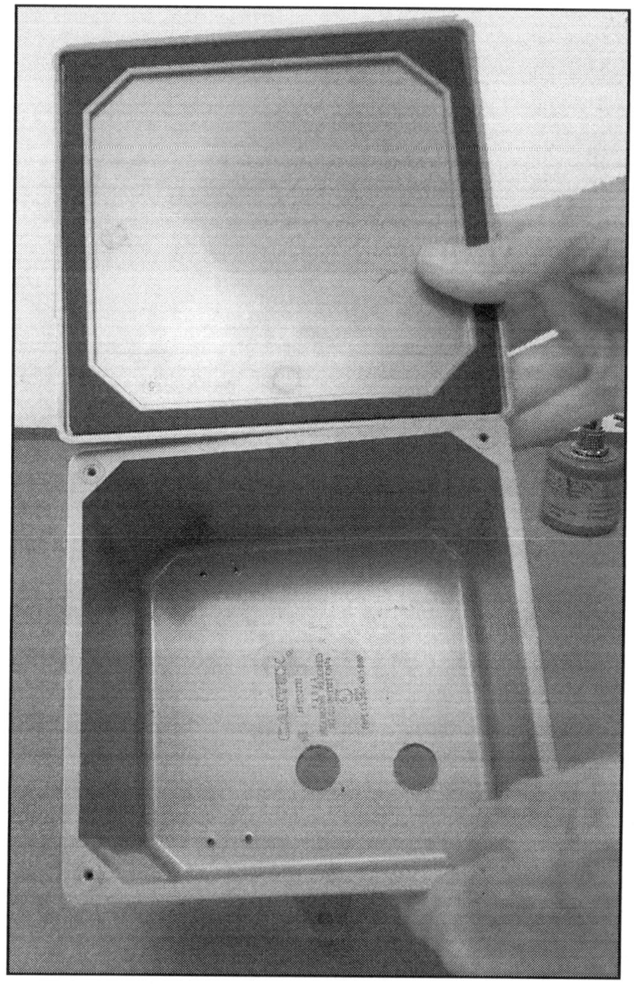

Ports and mounting holes for the terminal block were drilled in the box

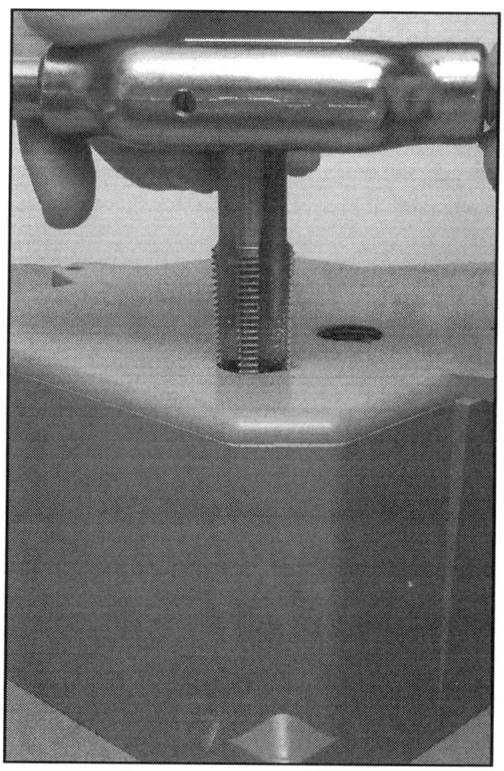

Tapping the port holes

one ground wire (connected to the lightning arrestor); and one UFB cable that connects the combiner box with the DC service box.

The location of the two ports was marked, as well as the location of four screws to attach the terminal block to the inside of the box. The inlet and outlet wire port holes were drilled with a 45/64 drill, then hand tapped with a ½" tap.

The four smaller holes were drilled for mounting the terminal blocks.

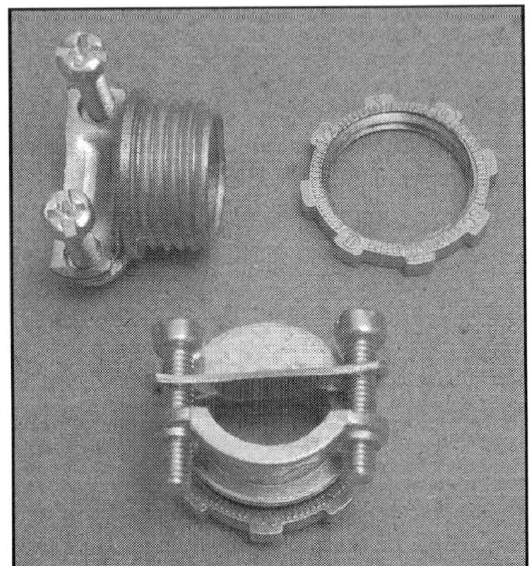

Then, the entry and output ports were fitted with strain reliefs (cord grips). The strain reliefs provide a flat surface for the wires to rest on and also prevent the wires from coming off the terminals if they are pulled from the outside. They also seal the ports from the elements to a certain degree.

There are several types strain relief available. We used the standard electrical die cast variety, but a better choice would have been a liquid tight or moisture tight type. After the combiner box was wired, the wire entry ports were covered with GE silicone II as a moisture barrier. Other types of strain reliefs are specifically made to deter the entry of

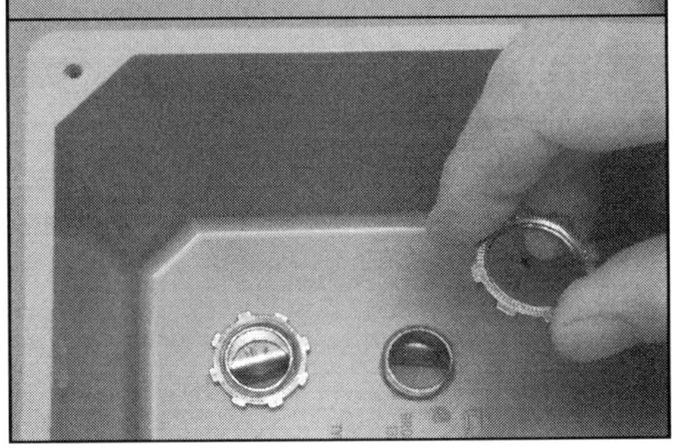

Strain reliefs were installed in the wire ports

moisture and liquid and can be found at suppliers such as McMaster Carr and solar goods suppliers.

When attaching the strain reliefs, coat the threads with silicone II so that a seal is formed to keep the fitting water tight. Any part of the fitting that you feel should be coated to seal from weather should be coated. Even with the water and moisture tight cord grips it pays to put some silicone on wire entry and exit ports after the wire has been installed.

After silicone was applied, the components of the strain relief were screwed together and tightened. With the simple die cast strain reliefs there are two screws on either side that have to be loosened to admit the wire. After the wire was inserted, the screws were tightened to partially cover the opening with a die-cast semi-circular shield.

Installing the terminal block

Once the strain reliefs were installed, the terminal block was screwed into place inside the box. Four jumpers were connected to terminals on the block (one jumper for each panel). Then, the combiner box was attached to the array rack.

Service entry

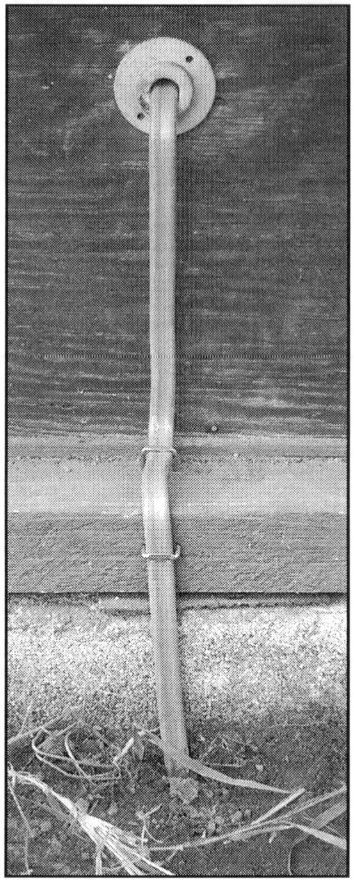

Completed service entry

The entry point of the wire run from the PV array to the structure where the power will be used will, of course, determine the location of the trench for the wire run. For this project, all the equipment would be installed in the basement of the structure being serviced. A basement window was removed and replaced with a sandwich of plywood and insulation. Holes were drilled for cable entry, ground wire, and vent exhaust pipe for the battery box.

For cable and wire ports, manufactured cable feed-throughs can be used. They are basically plastic tubes with end caps that attach to the outside and inside entry surfaces. The tube can be cut to length. These are sometimes called antenna feed-throughs. Most building supply stores will have a variety of service entry pipes and fixtures. Another option is to use PVC pipe. Use the smallest diameter pipe or feed-through that will do the job. All holes cut for ports should be as tight a fit as possible to the tube or pipe used.

Feed-through entries should be caulked with silicone around the outside surface of the pipe or tube that will be inside the wall, and between the edges of the hole in the wall and the tube on the outside and inside. This will seal the feed-through to the structure. After the cable is laid and threaded into the building through the tube, caulk inside the tube around the wire on the inside and outside of the wall to seal the opening where the wire comes through.

All wire should be flush to surfaces and out of the way so that people do not trip over them, or cut them with lawn mowers, etc.

Digging a trench for the wire run

The wire run from the solar panel to the DC service entrance is usually buried anywhere from 8" to 18" deep. The depth of the trench is dictated by the amount of traffic the surface ground bears. If there is a lot of foot traffic over the wire run, the cable should be buried deeper. If the wire run is through an area that has motor vehicle traffic, such as driveways, the trench needs to be 24" deep. Some areas have building codes that specify the depth for any incoming cable.

Trench for the wire run, with UFB cable laid in.

The wire run should be as straight a line as possible from the array to its entry point into the building.

If there may be more digging in the future for other projects near the buried wire run, boards can be laid on top of the cable along the run to prevent shovels from nicking the cable. Conduit can be used either above or below ground to further protect the cable. However, using conduit for underground burial, unless done very well, can create problems. If conduit is not completely sealed, liquid ingress can flood the conduit making a very bad electrical situation. I don't recommend using it unless you are sure that you can make it tight.

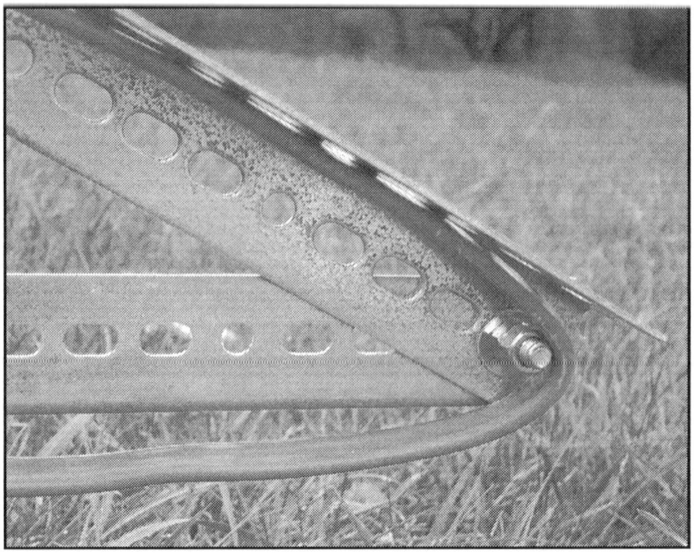

The UFB is routed to the front corner of the rack, then to the combiner box, so that the rack angle can be changed as needed.

The UFB coming out of the trench at the rack.

Laying the line

The cable used must be rated for direct burial. We used UFB. Always leave plenty of extra cable on either end when laying your line. Cable that's too long is a much, much better mistake than cable that's too short.

Wiring the Panel Array

Attaching the ground wire from the lightning arrestor to the rack

Installing the lightning arrestor

The lightning arrestor was placed in the combiner box. The arrestor red wire was connected to the positive terminal on the block, the arrestor black wire was connected to the negative terminal on the block. The arrestor green (ground) wire was threaded through the output port. (See photo next page, and wiring diagram for Terminal Block TBE on page 99.)

A ring terminal was crimped on the green ground wire coming from the lightning arrestor in the combiner box. The ground wire was inserted through the output wire port and attached it to the panel rack with a bolt and washers. It was tightly attached so that all the metal conductive surfaces have good contact with each other to create a good ground. Stainless steel bolts and washers were used to avoid corrosion. Once the assembly was tightened, it was coated with silicone II for added weather protection.

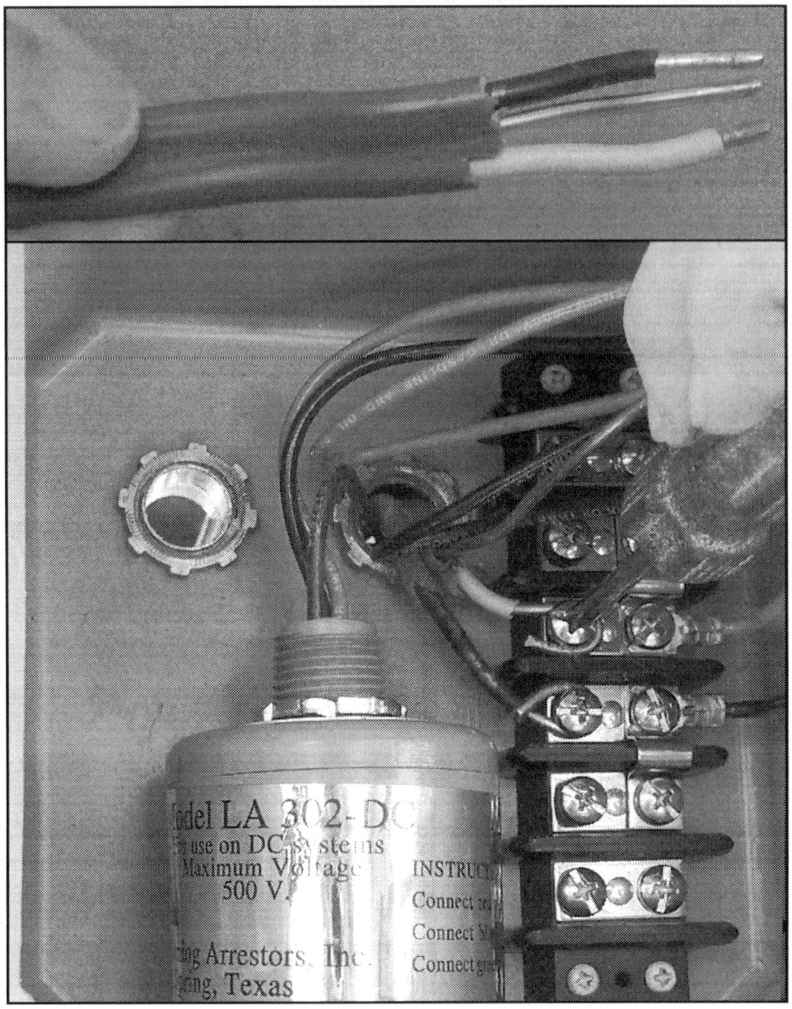

The lightning arrestor has been installed. Above, UFB wire with stripped end showing 3 wires. Below, the black and white wires of the UFB are attached to the block; the bare wire will be wrapped and not used.

UFB output

The UFB wire from the DC service box was routed through this same output port, and connected to the appropriate terminals. Note that what is referred to as two conductor UFB (commonly available), in fact has three conductors: one black sheathed wire, one white sheathed wire, and one bare wire (see photo, top). The bare wire is usually used as a ground wire.

The outside gray sheath was stripped off to expose the three wires inside the combiner box, but the cable enters the combiner box fully sheathed.

The UFB is 10 gauge wire which can be difficult to work with in tight spaces. The wires were shaped with needle nose pliers to fit the screw terminal, though terminals could have been placed on the ends of the wires for connection to the terminal block.

The wires were marked positive and negative and noted so that on the other end, at the DC service box, it would be clear which wire is which.

Since the bare ground wire was not used, it was covered with shrink tube. It could also have been cut off and the bare tip covered with silicone or liquid electrical tape. This bare wire should not be used for a positive or a negative because it does not have the insulation that the other two wires have.

After the wire was prepared to fit the terminal screws (or crimped on a terminal connector), the bare ends of the wires were wrapped with electrical tape since they would not be connected until the system was completely wired and ready to test. Wrapping the bare conductor ends keeps them from shorting out if the two leads touch each other, and prevents them from coming into contact with any conductive surface.

For additional weather protection, a drip line was bent into the wire outside the combiner box. A drip line is simply a curve in the wire positioned below the entry or exit port that lets precipitation roll off away from the box, rather than flowing down toward the opening in the box.

Connecting the panels

Four 12 volt panels that were constructed from the instructions in *Build Your Own Solar Panel* were used for the PV array. The four panels were connected together in parallel with 12 gauge red and black zip wire. The zip wire red was connected to the positive screw, and the zip wire black was connected to the negative screw on the terminal block inside the junction box of each panel. The lengths of zip wire needed were measured by using string. The zip wire would be threaded along the rack, so for each panel the measuring string was threaded from the junction box on the panel to the combiner box along the rack in exactly the same way that the wire would be, so that each wire length needed could be determined. This avoids cutting a piece of good wire too short and having to redo it. Always cut wire longer than you think you need, and then trim for the final connection.

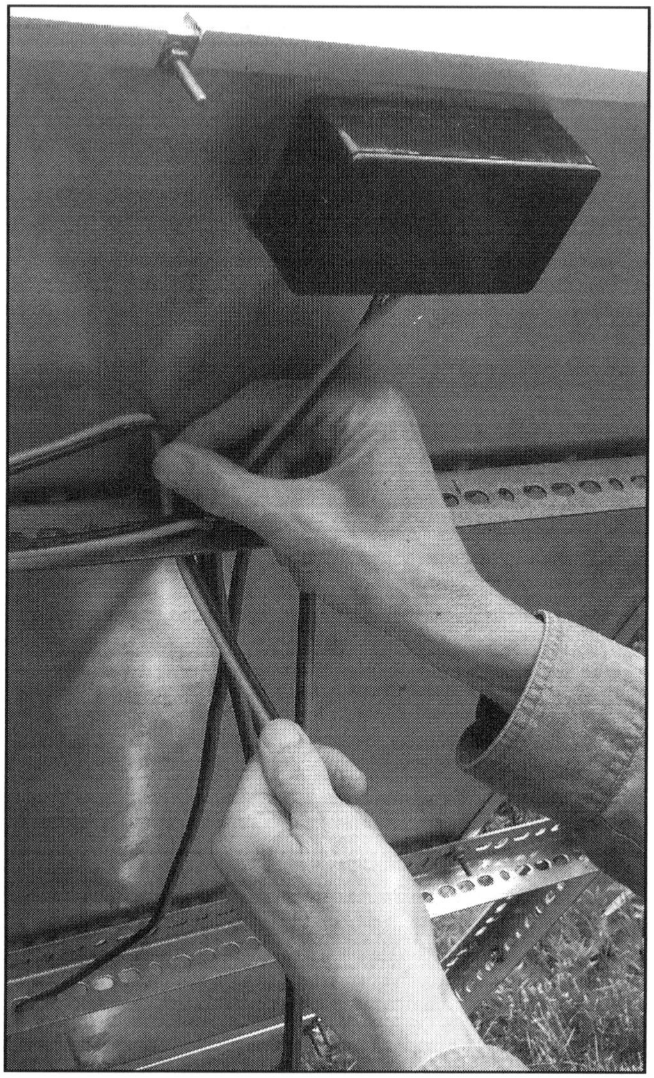

The zip wire was threaded along the rack to connect the panels to the combiner box. See photo next page also.

When cutting to final size, leave a little extra so that if a crimped terminal has to be cut off and replaced in the future because of corrosion, there will be enough wire to do so.

When fastening wires to screw terminals, you have the option to simply fasten the bare wire to the terminal, or attach spade or ring crimp connectors to the wire ends for placement on the terminals. The crimp connectors keep things neater, especially with stranded wire. For solid wire you can either use crimp terminals or bend the wire to conform to the terminal screw for attachment. (See section on wire terminals, p. 48.) The spade or ring crimp terminals will make the connections much easier to remove and replace when needed.

Zip cord is my choice to connect the panels because the ends can be separated in the box enclosures to connect to the combiner and junction terminals, but the wires can stay together along the run, which makes it neat and easy to install.

UV resistant tray cable could also be used for this application. Although the zip cord is handy to use, it is not rated for UV resistance. Since the zip cord is behind the panels, I don't consider it that much of an issue at our location; however if you have concerns about UV resistance, use tray cable.

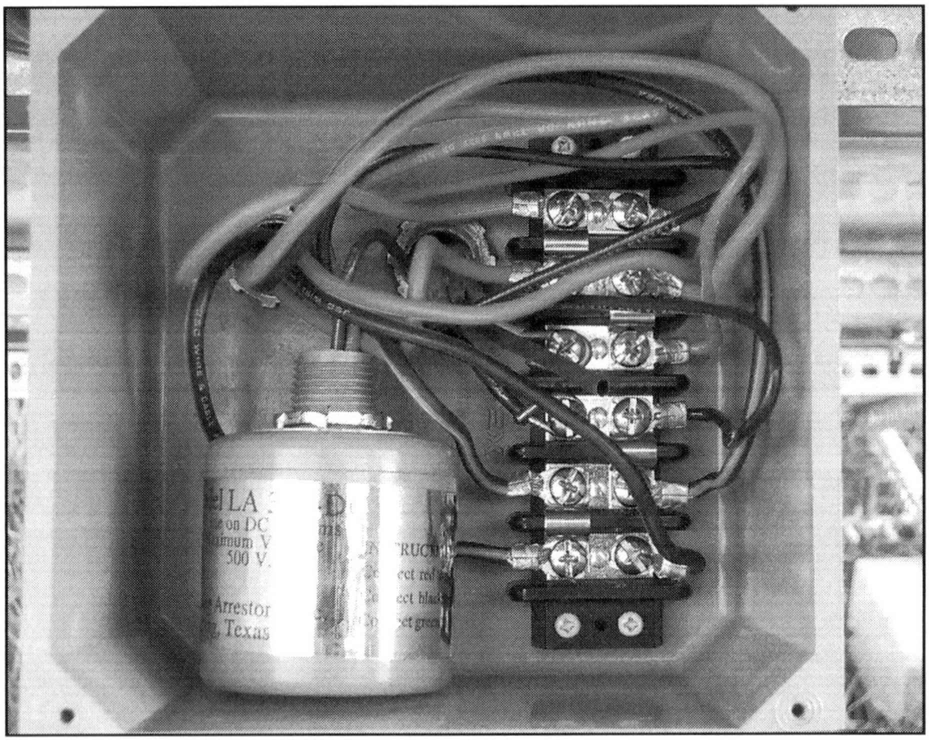

The completed array wiring in the combiner box

TBE - Terminal Block E

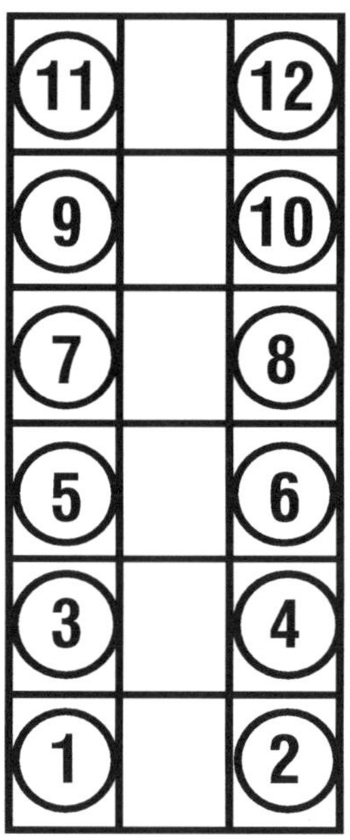

POSITION	CONNECTS TO
1	solar panel 1 negative, and position 3 on block via jumper
2	solar panel 2 negative
3	solar panel 3 negative, and position 1 on block via jumper
4	solar panel 4 negative, and position 6 on block via jumper
5	UFB negative
6	negative wire lightning arrestor, and position 4 on block via jumper
7	UFB positive
8	positive wire lightning arrestor, and position 10 on block via jumper
9	solar panel 1 positive, and position 11 on block via jumper
10	solar panel 2 positive
11	solar panel 3 positive, and position 9 on block via jumper
12	solar panel 4 positive

Completing the combiner box

Once all the wires were pulled and positioned in the box, the clamp screws on the die cast cord retainers were tightened. The exit holes were filled and covered with silicone to form a water and moisture tight seal. Other types of cord retainers will seal in different ways. With some there may not be a need to add silicone if the hole is compression tight, but it doesn't hurt to add a little more protection by adding silicone.

Tools

45/64" drill
½" NPT tap
Electric hand drill or drill press
Caulking gun
Hand tap for ½" NPT size tap

Parts for the combiner box

ITEM	SOURCE	PART NUMBER	HOW MANY
Junction box, 6"x 6"x 4" with gasket	Cantex Inc., McMaster Carr, or hardware store	Cantex 5133710	1
Terminal Block E - 300 VAC/VDC terminal block 6 circuits, .69" center-to center, 65 amps	McMaster Carr	7527K26	1
Jumpers for .69" center-to center 300 VAC/VDC terminal block (also listed for DC service box)	McMaster Carr	7527K39	1 pack of 25
Lightning arrestor	Backwoods Solar Electric Systems, or Delta	LA-302DC	1
Silicone caulk	Hardware store		1 tube
Assorted screws or bolts, and nuts and washers to fasten box and grounding to panel rack	Hardware store		
Metal or plastic strain relief/cord grip	Hardware store, McMaster Carr, solar components distributors		2

DC Service/Distribution Box

The DC service distribution box for the Solar II project is the focal point for the incoming DC from the solar array. It houses the charge controllers, various switch controls, fan control, voltage readout, LED indicators, fuses, and distribution blocks.

Enclosure

An enclosure specifically for this purpose can be purchased from a solar component distributor; or, other retail or surplus suppliers will probably have something that can be adapted for your needs.

Several features should be considered. Most important is the number of knockouts or port holes available. The box should have at least the minimum of holes needed for wire entry and exit according to your design.

The enclosure should be metal and not plastic. Plastic can burn if exposed to shorts in wiring or high heat from some components.

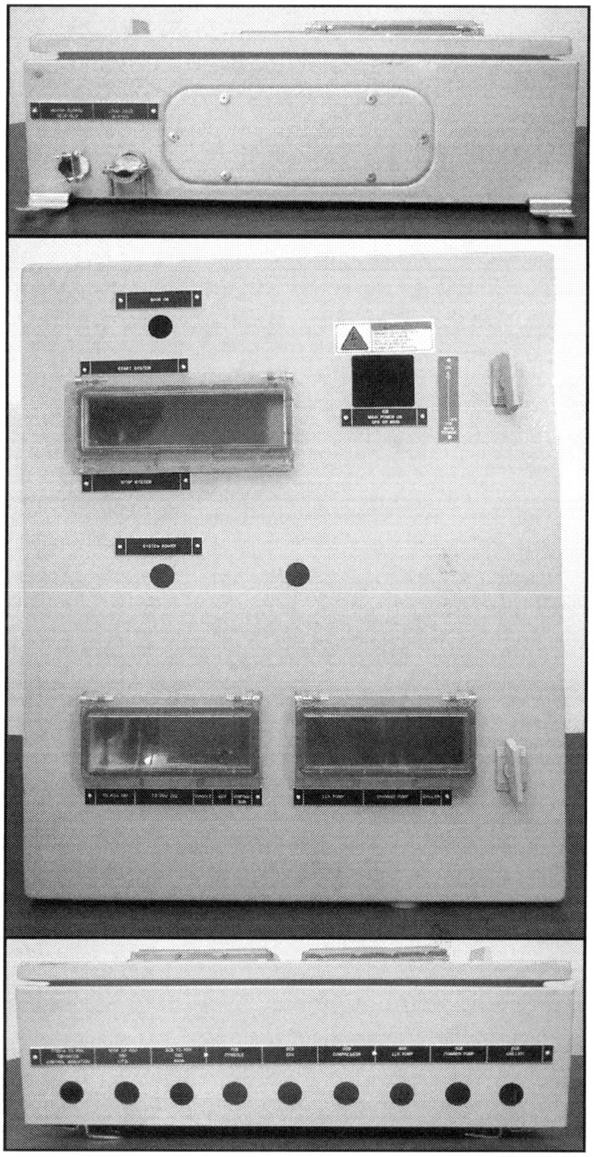

Top, front, and bottom views of the box used

If the enclosure will house components that give off great amounts of heat, there must be sufficient openings for convective cooling.

In most cases, convective cooling is adequate, but you may also want to use a muffin fan to pull air through the box if you wish to cool the components further. Most circuits do not require much ventilation, but some circuits, such as the transistor based charge controller (CC2), must have sufficient ventilation to keep the component temperature down to avoid device failure.

The particular box used for this project was much larger than normally necessary. The dimensions are 19½" wide x 25¼" high x 5½" deep. We wanted a very large box in order to experiment with several different types of charge controllers, have room for additions, and have the capability to rework any of the components if desired.

Finding the right enclosure took some time. There were many sources of both new or used. This particular box was being offered on an internet auction site at a price that we could not walk away from, considering its features.

It had a removable plate on top, and many cable entry ports. The front door has various holes with plastic windows that open and close, two front latches and a gasketed door. These features made it easy to fit in components such as an LCD readout, LED indicator lights, and switches.

There are also mounting brackets on the outside, and a steel platform for components inside that attached to four bolts situated in the corners of the box. The bolts had plastic standoffs so that the component board was raised from the back surface, making space between the platform and the back of the box.

The steel component platform was removed and replaced with a 1⁄16" thick G10/FR-4 board which was more lightweight. A 24"x 24" piece of FR-4 was purchased and cut to fit the box dimensions.

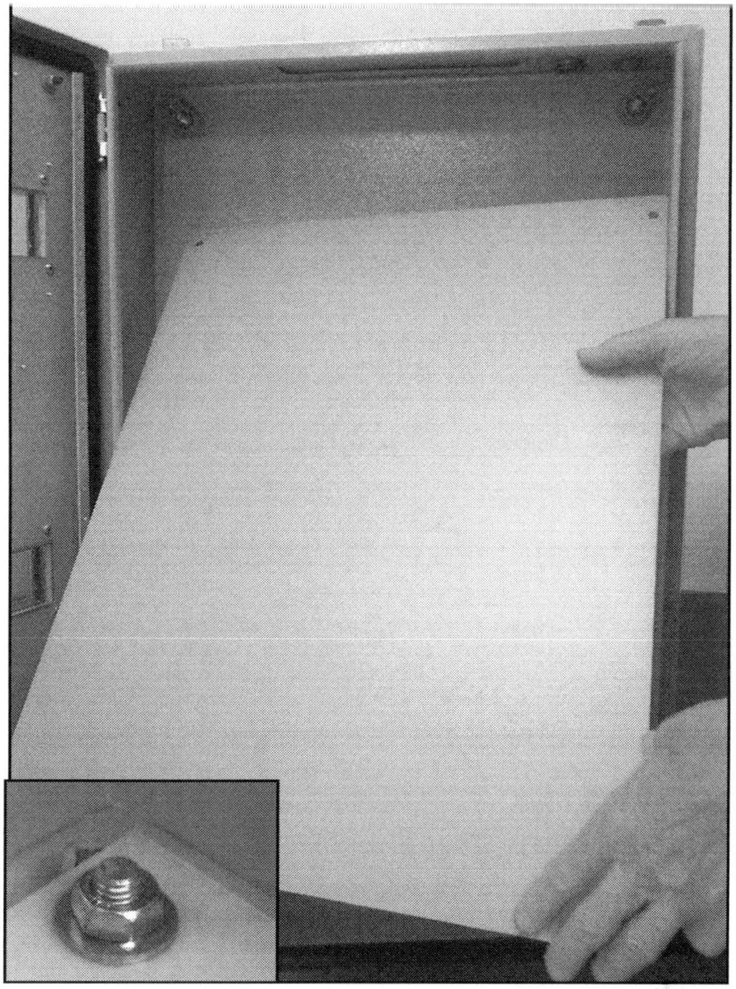

Holes were then drilled in the FR-4 to match the bolts' positions in the box (see photo at right). This created a handy insulated circuit platform with space to route wires behind the board.

Although the steel plate would have worked for a circuit platform, we wanted an insulating material, that would be light and easy to drill and cut. Since this was an experimental circuit board that would have changes made in the future, the FR-4 was a good choice.

G10/FR-4 is flame retardant, with a maximum operating temperature of 284°F. The 1⁄16" thickness of this board was adequate, but the span of the board caused it to sag at the center when the heavier components were placed. Plastic standoffs were placed and glued at various points on the back of the board for support. Another option for wide spans would be to use a thicker board such as 1⁄8".

The G10/FR-4 is the circuit platform for a variety of components, including two types of charge controllers, exhaust fan control for the battery box, digital voltage read out, and switch controls for cutting off power and switching between two types of charge controllers. The box also contains two fuses. One fuse is on the connection from the charge controller to the battery bank positive, and another is on the positive wire coming from the batteries to the inverter.

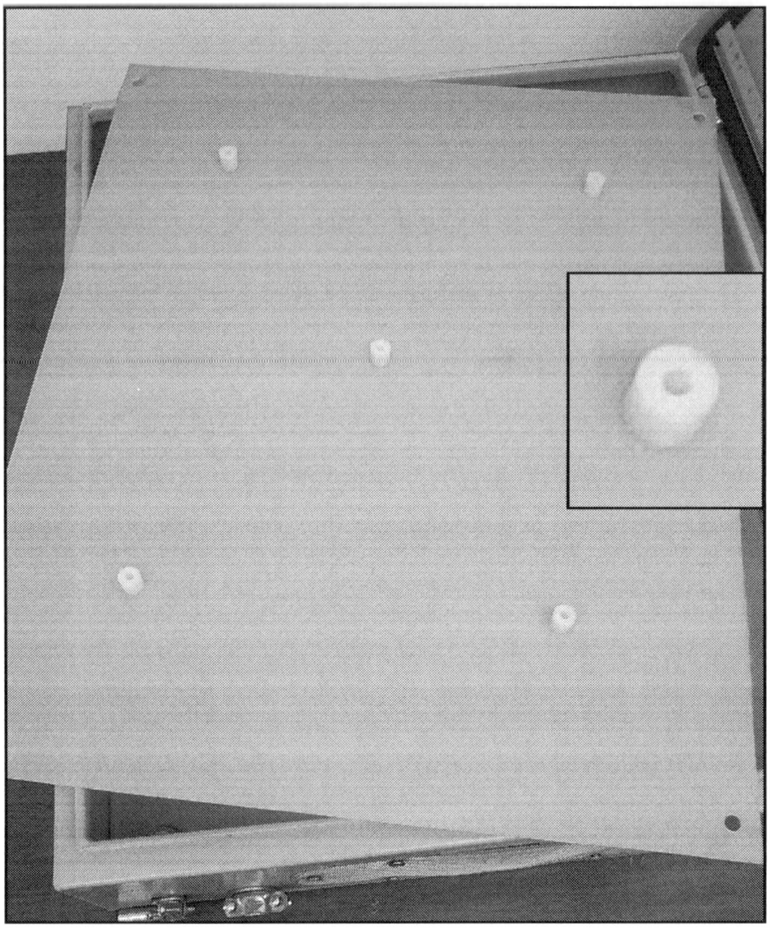

Plastic stand-offs were glued to the back of the G10/FR-4 sheet.

Fuse between batteries and inverter

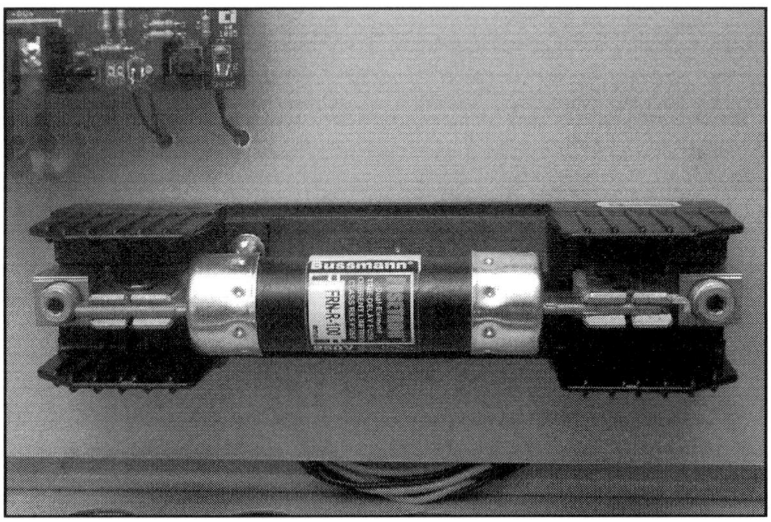

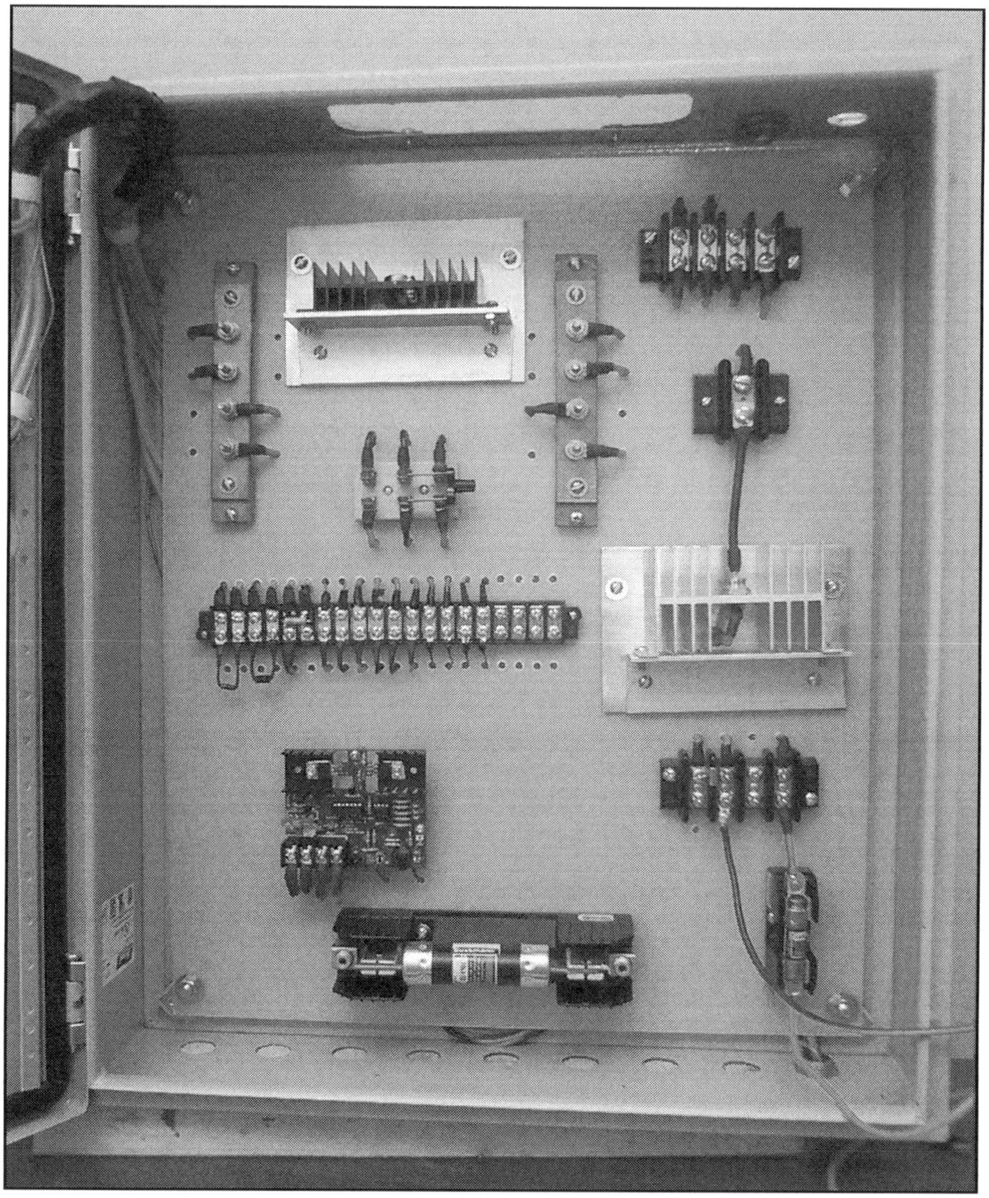

Layout for components in DC service box

To plan the layout of the DC service box, you have to know exactly what the circuit will be and have the components available for test placement and layout to see if everything fits correctly in the box. Prior to attaching the components to the case and board, several different configurations on the FR-4 board and case were considered for component placement.

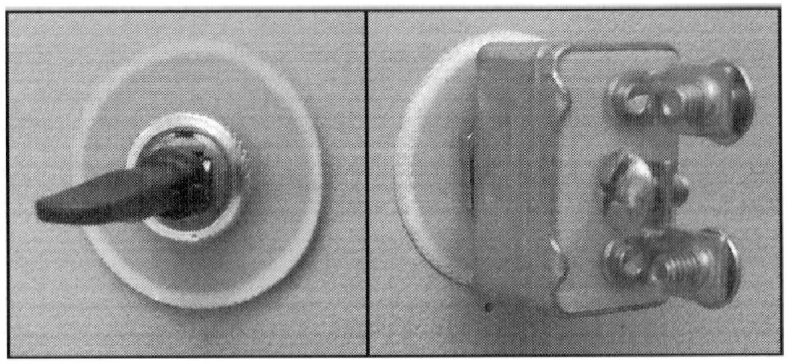

Toggle switch, outside door, left; inside door, right

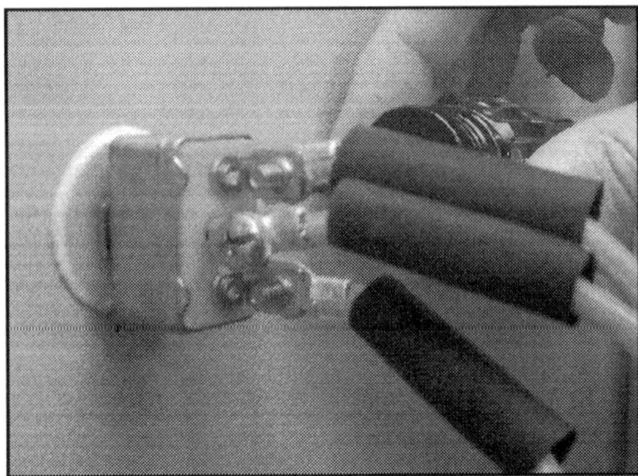

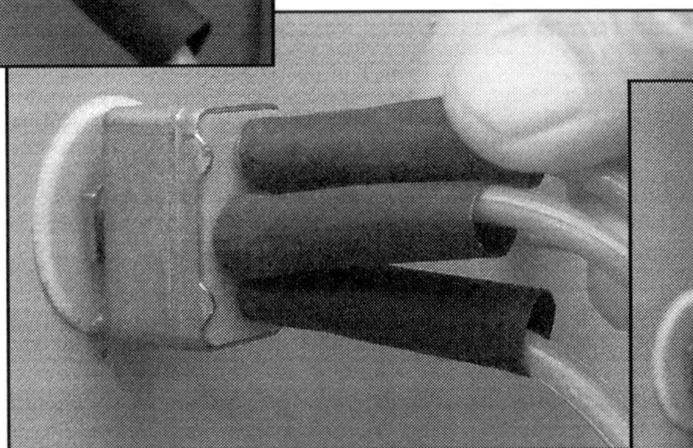

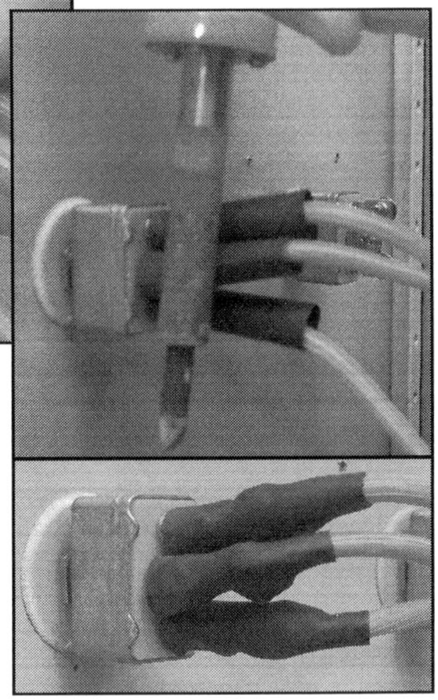

Connections to the switch were covered with heat shrink tubing

Once the component placement positions were determined, the components were placed on the board and the screw holes for mounting each component were marked, and the attachment holes drilled in the FR-4. This included holes for two bus bars, two heat sinks, four terminal blocks, two fuse holders, and the kit circuit charger board.

When considering component placement, components that will be connected to external wire should be located near ports so that heavy input and output cables do not have to be twisted and bent excessively.

If a box front will contain LEDs, switches or meters, make the appropriate holes for the components or their fastening holes.

Left, face of battery vent switch; and below, installing the switch on the inside of the box door.

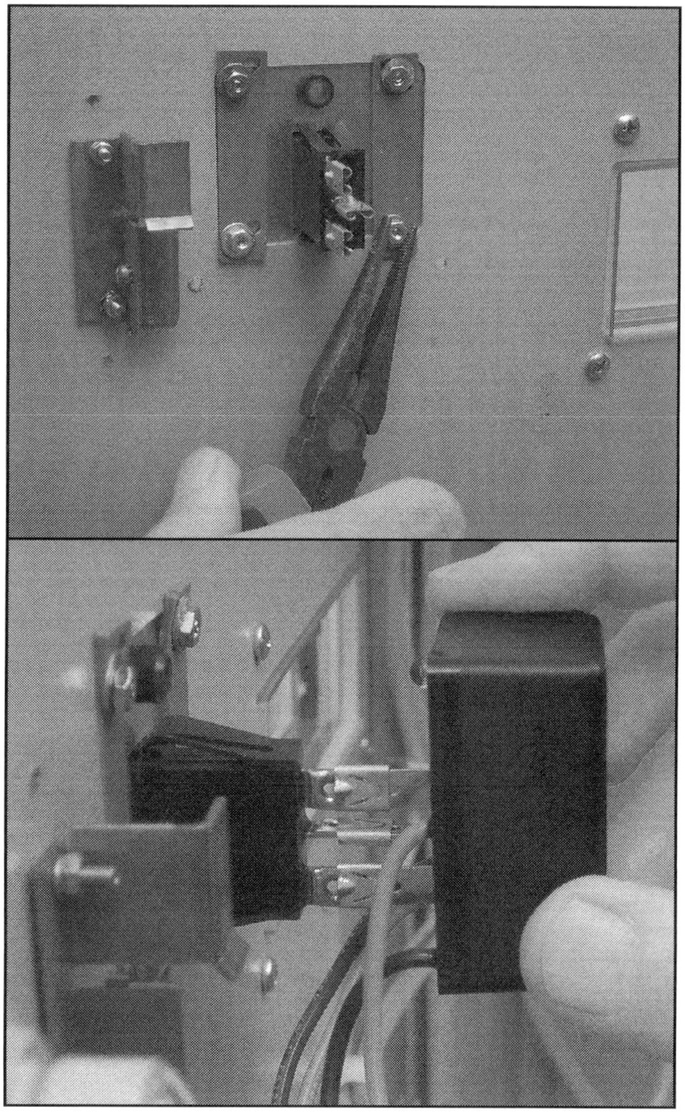

Components mounted on the face of the box require good wire management by using wire holders and tape to keep the wires from interfering with opening and closing the box. The wires should be protected so that opening and closing the box will not abrade them. Also, no routed wires should be near or touch heat sinks, or lie on or interfere with other components on the board when the box front is closed. Components placed on the front must not be not too close to heat sinks. In general, allow as much room between components as possible.

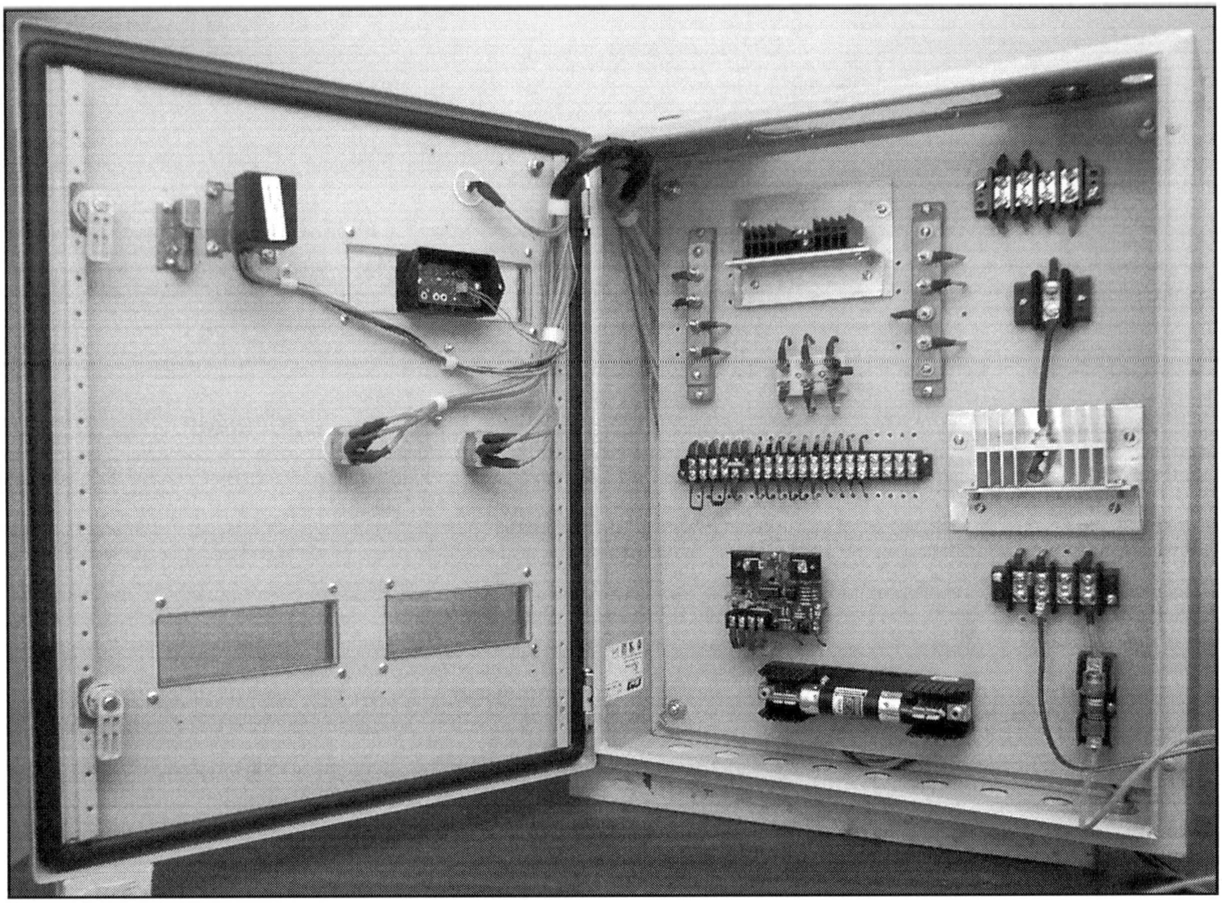

The DC service/distribution box with all components installed

For this DC service box, most of the wires were routed behind the FR-4 so that they would not be on the surface where they would be in the way and a source of entanglement. This is not the only way to do it, but if you do wire on the surface, use wire stays to keep it neat, and keep the wires close to the surface so that they do not catch on fingers and components.

The UFB cable was routed to the inside of the box from the top right port. A cable grip was inserted to secure the cable. Before the UFB was connected, the bare ground wire, which was not being used, was trimmed and coated with liquid electrical tape to insulate it. Crimp on terminals were added to the negative and positive wire

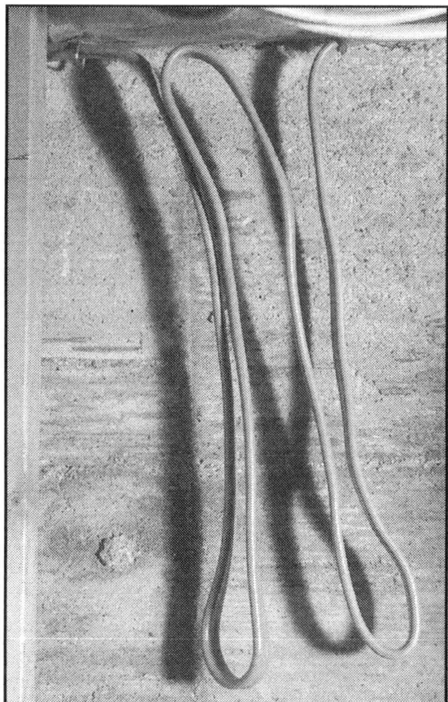

The excess length of the UFB cable coming from the panels was "folded" before entering the DC service/ distribution box

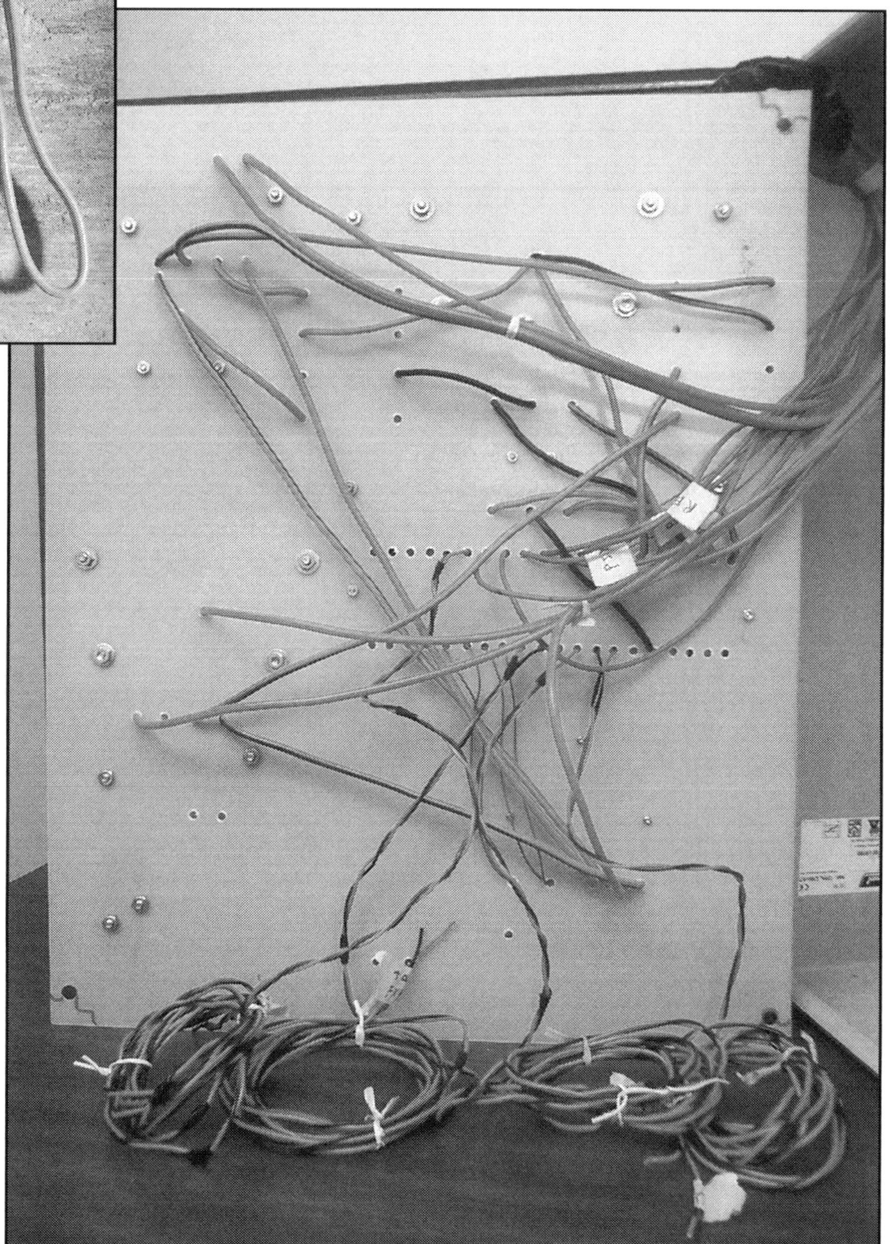

Wiring routed behind the FR-4 sheet in the DC service/ distribution box

of the UFB and insulated with shrink tube. The incoming positive and negative wire of the UFB was connected to a four position, 30 amp terminal block with eight screw connectors (TBA).

TBA – Terminal Block A

POSITION	CONNECTS TO
1	UFB negative
2	CC2 via negative bus bar BB2A CC1 negative input
3	UFB positive
4	positive to S1 pole
5	S1 throw 1
6	CC1 positive input
7	S1 throw 2
8	CC2 via positive bus bar BB1A

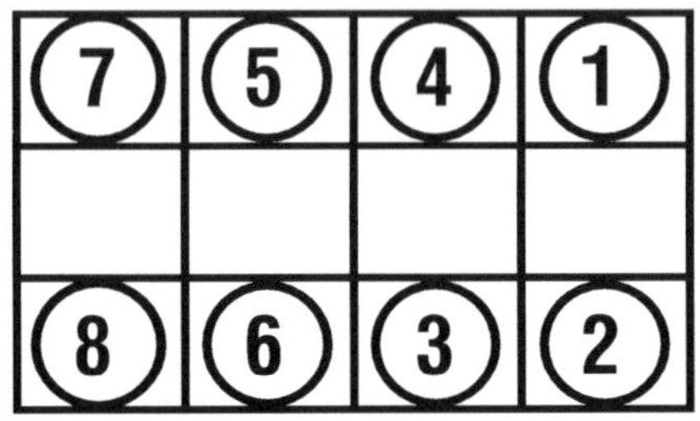

CC1 – charge controller 1
CC2 – charge controller 2
S1 – single pole double throw switch on the box door (on, off, on positions)
UFB – connects to solar array

The single pole double throw switch in the middle or off position cuts off the positive current from the panel to the box. Throw one allows the positive current of the panel to proceed to CC1. Throw two allows the positive current to flow through to CC2. This particular switch was rated at 50 amps and had screw terminals for wire connections. Each screw terminal connection was covered with heat shrink tubing to insure good electrical insulation from other wires and components.

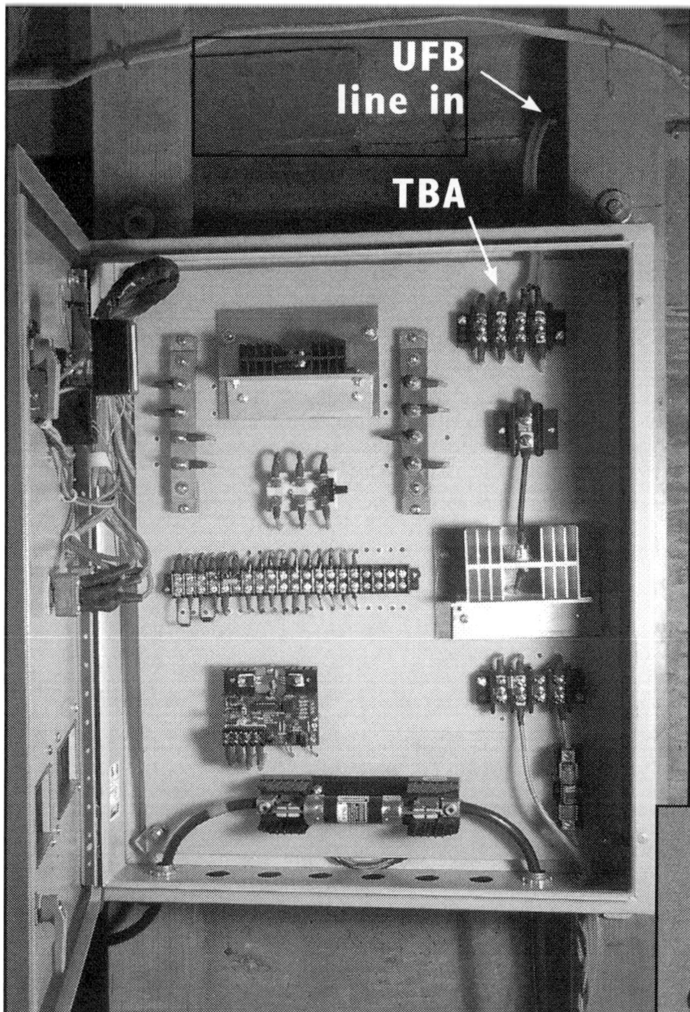

All of these connections were made with #12 zip cord with crimp terminals and heat shrink tube insulation. All wires were routed behind the FR-4, so appropriate wire holes had to be drilled in the FR-4 to accept the wires. Wires to the cabinet switch were routed from under the FR-4 along a wire bundle situated at the left top of the box.

In addition to CC1 and CC2 charge controllers, the DC service box contained a digital voltmeter, and a fan switch for the battery box which automatically exhausts hydrogen gas, and electrolyte fumes outdoors as the batteries gas.

Digital voltmeter, front and back

The completed DC service/ distribution box

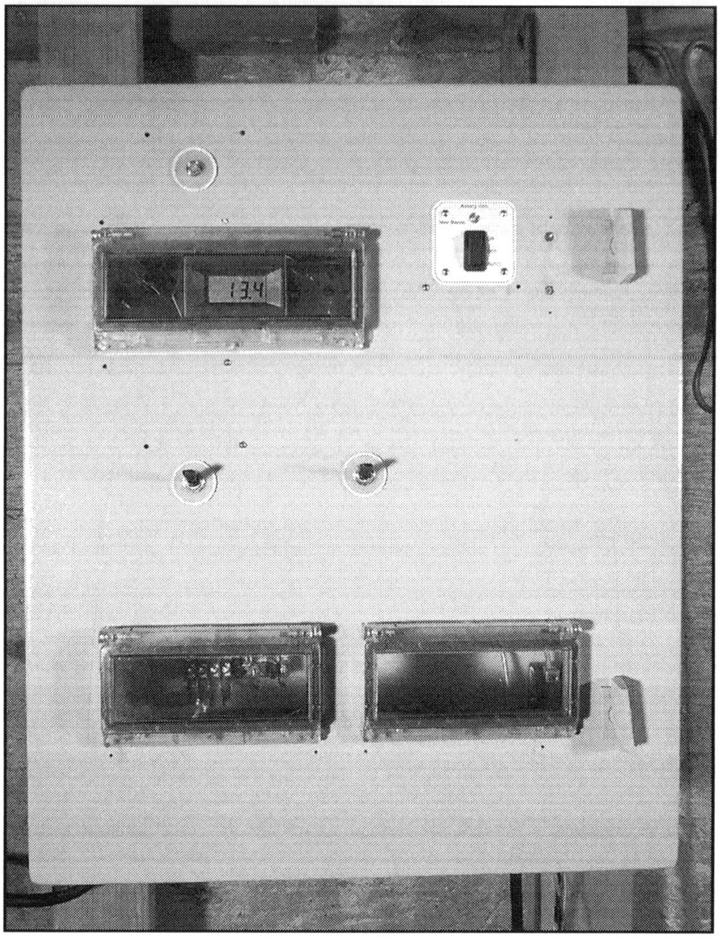

Tools for the DC service/ distribution box

Saw

Screwdriver

Pliers

Drill and assorted drill bits

Parts for the DC service/distribution box

ITEM	SOURCE	PART#	HOW MANY
Terminal block A (TBA), Terminal block D (TBD) – 300 VAC/VDC terminal block 4 circuits, .69" center-to-center, 65 amps	McMaster-Carr	7527K24	2
Terminal block B (TBB) – 300 VAC/VDC terminal block 20 circuits, .44" center-to-center, 20 amps	McMaster-Carr	7527K61	1

DC Service/Distribution Box

ITEM	SOURCE	PART#	HOW MANY
Jumpers for TBD, .69" center-to-center 300 VAC/VDC terminal block	McMaster-Carr	7527K39	1 pack of 25
DC service box	Surplus or new, many suppliers		
100 Amp class R fuse holder with two fuses	Backwoods Solar	O-FB100	1 set
30 Amp class R fuse holder with two fuses	Backwoods Solar	O-FB30	1 set
LED holder for CC1 LED	Electronics supplier		1
Digital voltmeter	Backwoods Solar	M-DIG Voltmeter	1
Toggle switch (50 amp DC)	Backwoods Solar	O- Toggle	2
Grade G10/FR-4 Garolite sheet 1/16" thick, 24"x 24"	McMaster Carr		1
Assorted screws, nuts, bolts, washers for fastening	Hardware store		
1 amp fuse for digital voltmeter	Electronics supplier		1
Inline fuse holder for voltmeter fuse	Electronics supplier		1
Bus bars	See bus bar section		
Exhaust fan control switch with fuse and inline fuse holder	See battery vent parts list		

Bus Bars

The CC2 circuit (charge controller) has solid copper bus bars for some connections. Bus bars are simple electrical connectors made from copper sheet, bar, braid and tube with studs, screws or other type of fasteners to hold connecting wires. They provide a low resistance current path for a variety of electrical connections in a circuit.

Bus bars can be used as terminals for general electrical connections. They are easy to rewire, so they're good in experimental systems that will be changed from time to time. Bus bars are frequently used in battery connections for larger systems, combiner box connections, and grounding blocks.

You can either purchase bus bars from commercial suppliers or make your own. When making bus bars, I design them to carry more current than I immediately need, so that if I expand a system in the future the same bus bar can carry the extra current produced by additional solar panels. As a general rule bus bars should be designed to carry anywhere from 30% to 150% more current than needed for a particular project.

Calculating current carrying capacity

The measuring unit for the current carrying capacity of a bus bar is the square mil. One square mil equals .001". So, to calculate the current carrying capacity of a bus bar made of .125" thick copper rectangle that is .5" wide, move the decimal 3 places to the right, which is 125 and 500. Then multiply these numbers to get the square mils, which in this case is 62,500. Refer to the table below, which indicates that the current carrying capacity of this copper rectangle will be somewhere between 125 to 150 amps.

Square Mils	Current carrying capacity/amps
3,226	20
5,128	30
8,152	35
12,967	50
20,617	70
32,783	90
52,127	125
65,730	150
82,860	200

Protecting bus bars

All bus bars must be enclosed, whether used indoors or outdoors. If a bus bar will be used in outdoor service it should be tinned which helps to prevent oxidation. This is easily done with an electroless tin plating solution. Another option to prevent oxidation is to coat the bus bar with a plastic spray or paint-on plastic (after making the connections). Bus bars used for indoor service can be either tinned, or coated or not at your discretion. Bus bars that will be near batteries must be protected from corrosive fumes by plating or coating.

Bus bars can be made to provide any number of connections. For this DC service box we made two bus bars with 4 connections on each bar.

Bus Bar Construction

For the DC service box, two 5.5" long bus bars were constructed from 1" wide, ⅛" thick copper bar, 10-24, screws and nuts, #10 copper washers, and ⅜" thick GP03 electrical grade fiberglass. Copper is the best material to use for bus bars as it is highly conductive. GP03 is great for a surface mount because it is a very good insulator.

Construction was fairly simple.

Preparing the parts

Copper bars

1. Copper bars were cut to two 5½" lengths with a hacksaw and miter box. (A bandsaw or chop saw with a metal cutting blade could be used for this.)
2. Each bus bar was marked at ⅜", 1¼", 2¼", 3¼", 4¼", 5⅛".
3. Pilot holes were drilled so that the drill would not slide when starting to drill the final holes. The pilot hole is simply a surface break to center the final drill point accurately. The pilot hole is drilled with a very small drill bit.

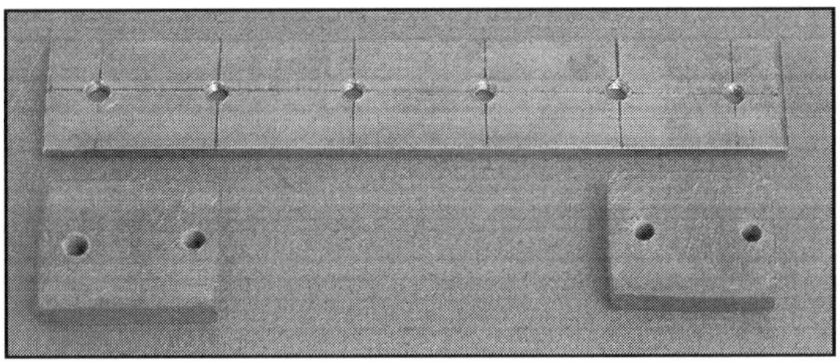

Copper bar and fiberglass blocks for one bus bar

4. The final terminal screw holes were drilled with a variable speed hand drill at a slow rpm and a #26 bit. Drilling at a low speed keeps the metal from hardening and becoming brittle. This made subsequent tapping easier and produces a better thread.

5. After cutting the bar, drilling holes, a hand file was used to de-burr and remove any ragged edges at the cuts.
6. The holes were tapped with a 10-24 hand tap.
7. Surface imperfections such as scratches and oxidation were removed by sanding with various grits of sandpaper, and final finishing with a Teflon scouring pad or fine sandpaper.

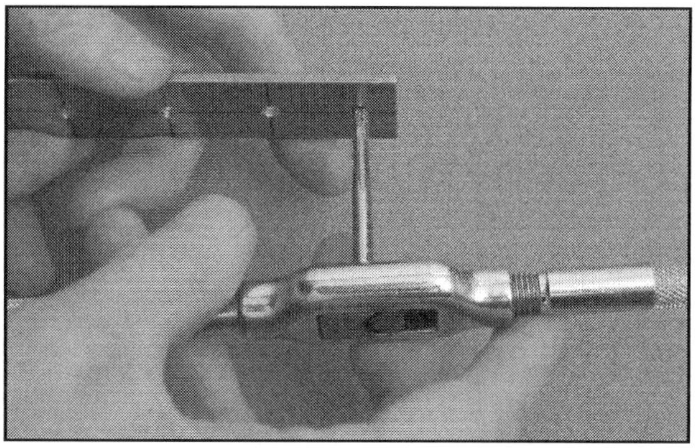

Tapping the holes

Fiberglass blocks

1. Four 1"x 1¼" blocks were cut from ⅜" thick GP03 fiberglass sheet.
2. A center line was marked down the 1¼" length of each block. Two points for holes were marked on the center line of each block, ¼" from each edge of the block.
3. Pilot holes were drilled at the marked points, then a #26 drill was used to drill the final holes.
4. The holes were tapped with a 10-24 tap.
5. The blocks were sanded to remove imperfections.

Assembly

Each bus bar terminal is composed of 4 terminal connectors. Each terminal connector consists of one 1" 10-24 screw, two 10-24 nuts, and four #10 copper washers.

Terminal connector

Each terminal bar is connected to each GP03 surface mount insulator with two ½" 10-24 screws and two #10 copper washers. The surface mount fasteners consist of two ¾" 10-24 screws that are inserted through the GP03 and fastened by a 10-24 nut and washer to the FR-4 sheet.

1. All the metal parts were cleaned by wiping the part with alcohol or other solvent, and then rinsing in clean water. Non-metal parts were cleaned in water.
2. Th bus bars were assembled by connecting terminals and GP03 surface mount insulator.

The ½" 10-24 screws that attach the copper bar to the GP03 block come close to the edge of the insulating block, but do not protrude from the GP03 with the copper washer in place. This small gap was filled with a dab of epoxy or liquid rubber to provide some insulation for a margin of safety. This protects from short circuiting that

Completed bus bar: top, side and bottom views

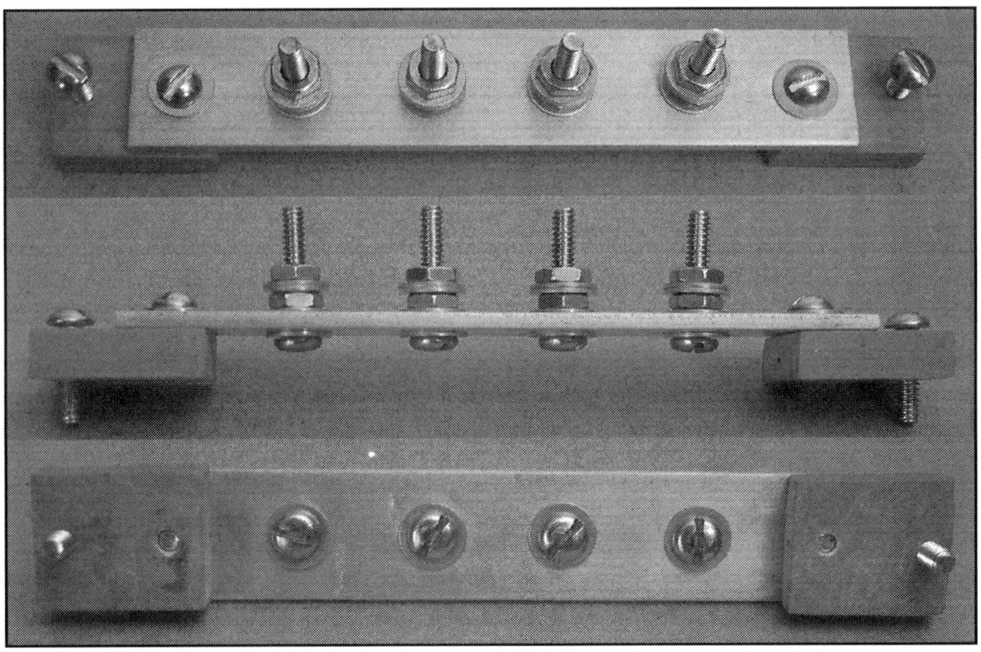

could occur if this terminal were mounted to a metal box surface and there happened to be a small bit of metal filing or wire clipping that contacted the screw and the box.

Tools for constructing bus bars

TOOL	SUPPLIER	PART #
Cobalt steel jobbers' twist drill bit heavy duty, wire gauge size 26, 3" L, 1⅞" L flute	McMaster Carr	3033A238
Machine screw size high-speed steel hand tap taper, 10-24, H3 pitch diameter, 4 flute	McMaster Carr	2522A671
Drill		
Screwdriver		
Pliers		
Hacksaw or cutoff saw		
Tap wrench straight handle style, 0-½" (1.6-2.5mm) tap cap	McMaster Carr	25605A75

Materials for constructing bus bars

ITEM	SOURCE	PART #	HOW MANY
Alloy 110 copper rectangle ⅛" thick, 1" width, 3 ft. length	McMaster Carr		One 3 ft. length
Copper flat washers, #10 screw size	McMaster Carr		50 per package
Electrical-grade fiberglass sheet (GPO3) ⅜" thick, 12"x 12", red	McMaster Carr	8549K47	One 12"x 12" sheet
18-8 ss pan head slotted machine screw 10-24 thread, 1" length	McMaster Carr		Packs of 100
18-8 ss pan head slotted machine screw 12-24 thread, ½" length	McMaster Carr		Packs of 100
18-8 ss pan head slotted machine screw 12-24 thread, ¾" length	McMaster Carr		Packs of 100
18-8 ss machine screw nut 10-24 screw size, ⅜" width, ⅛" height	McMaster Carr		Packs of 100
Commercial bus bars (optional)	McMaster Carr	see catalog	
Tinnit electroless tin plating solution	Electronics supplier	ER-18	1

Charge Controllers

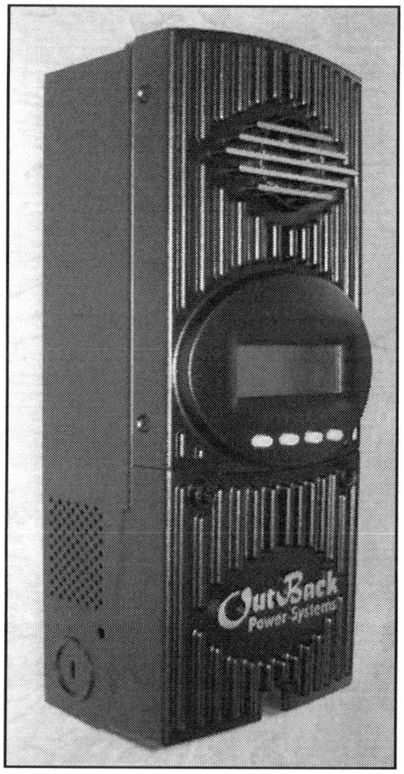

photos courtesy
Independent Power & Light

Charge controllers regulate the flow of electricity from the PV panels to the batteries. Their circuits are designed to keep the batteries fully charged and avoid overcharging and undercharging.

A charge controller can either be simple series, or shunt, which disconnects, shunts (diverts) the current to another circuit, or short circuits the array; or PWM (pulse width modulation) which apply a pulse to charge the batteries.

The PWM types are the most efficient and are purported to prolong battery life and prevent sulfation of the batteries.

Some chargers also have maximum power point tracking. This allows the panels to deliver maximum power electronically and is particularly useful in colder regions with long winters and much cloudiness, to make the most of whatever charge time is available.

There are commercial charge controllers that can be purchased; or they can be built either from a kit or plans.

Our DC service box has two charge controllers. CC1 is a PWM type which was constructed from a kit; and CC2 is a single transistor switch shunt controller that was constructed from parts easily available from electronics parts suppliers. CC2, although less sophisticated and less efficient for most applications, has a special purpose. A future project will be to integrate one of several electrolyzers into this PV system for hydrogen production. The CC2 circuit with minor modifications is more compatible with the requirements of the hydrogen system than CC1, as well as being able to act as a charge control for the batteries.

Either one of these charge controllers could be used alone in a system or both could be included, one being used as a backup when another one fails. Both are designed for a 12 volt system with no more than 20 amps maximum current input.

A thermistor mounted on one of the batteries is connected to CC1 for temperature compensated charging.

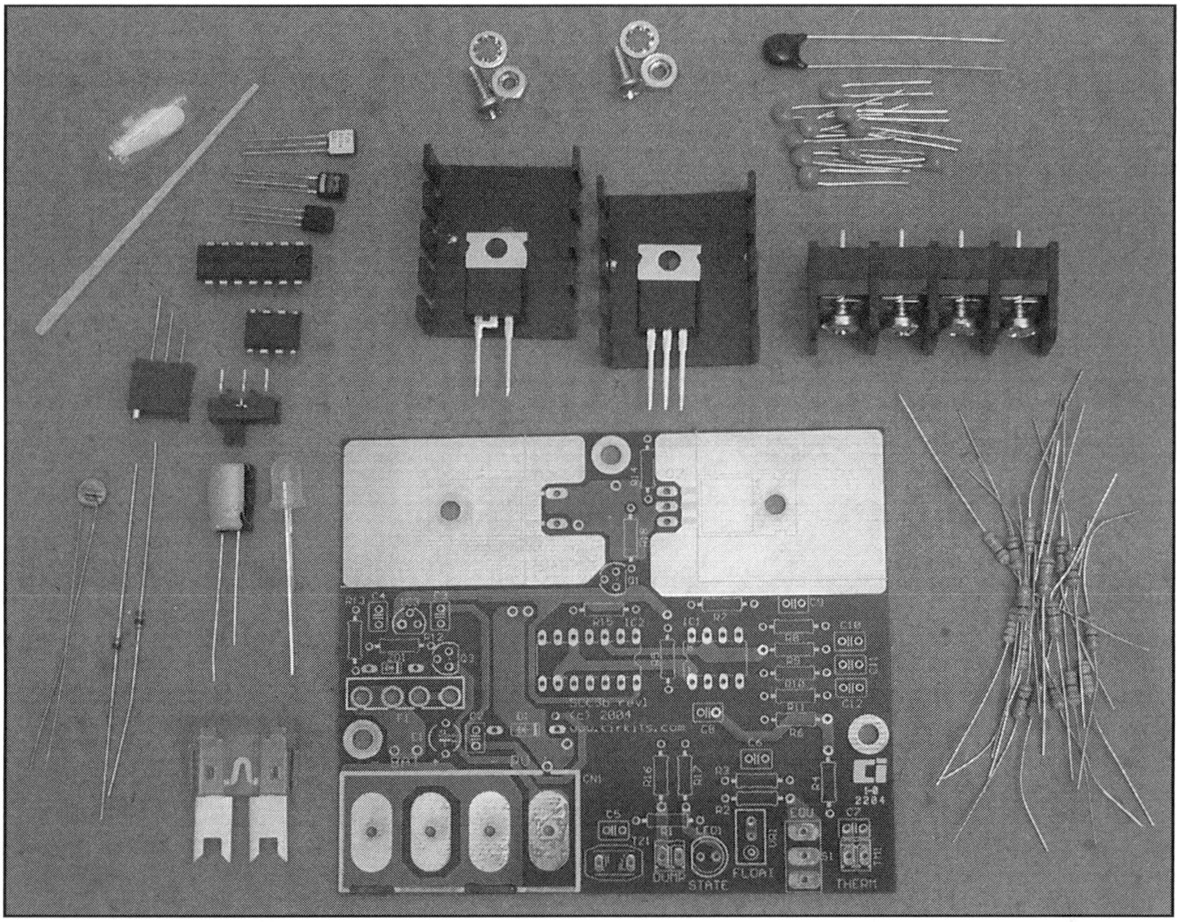

CC1 – pulse width modulation charge controller

CC1 is a pulse width modulation charger, meaning it delivers the charging current in pulses. It has a thermistor that can be attached to a battery surface for temperature compensated charging, and an LED charge indicator that gives an excellent visual of the charging process. Preferred charging voltage can be set, and it has a manual equalizing charge switch. This charge controller was assembled from a kit and comes with a 3"x 3½" printed circuit board for mounting the other components. The kit was easy to assemble,

comes with excellent instructions for assembly and use, and is reasonably priced. It is suitable for both flooded and sealed lead acid batteries. We highly recommend this kit. See page 181 for where to purchase.

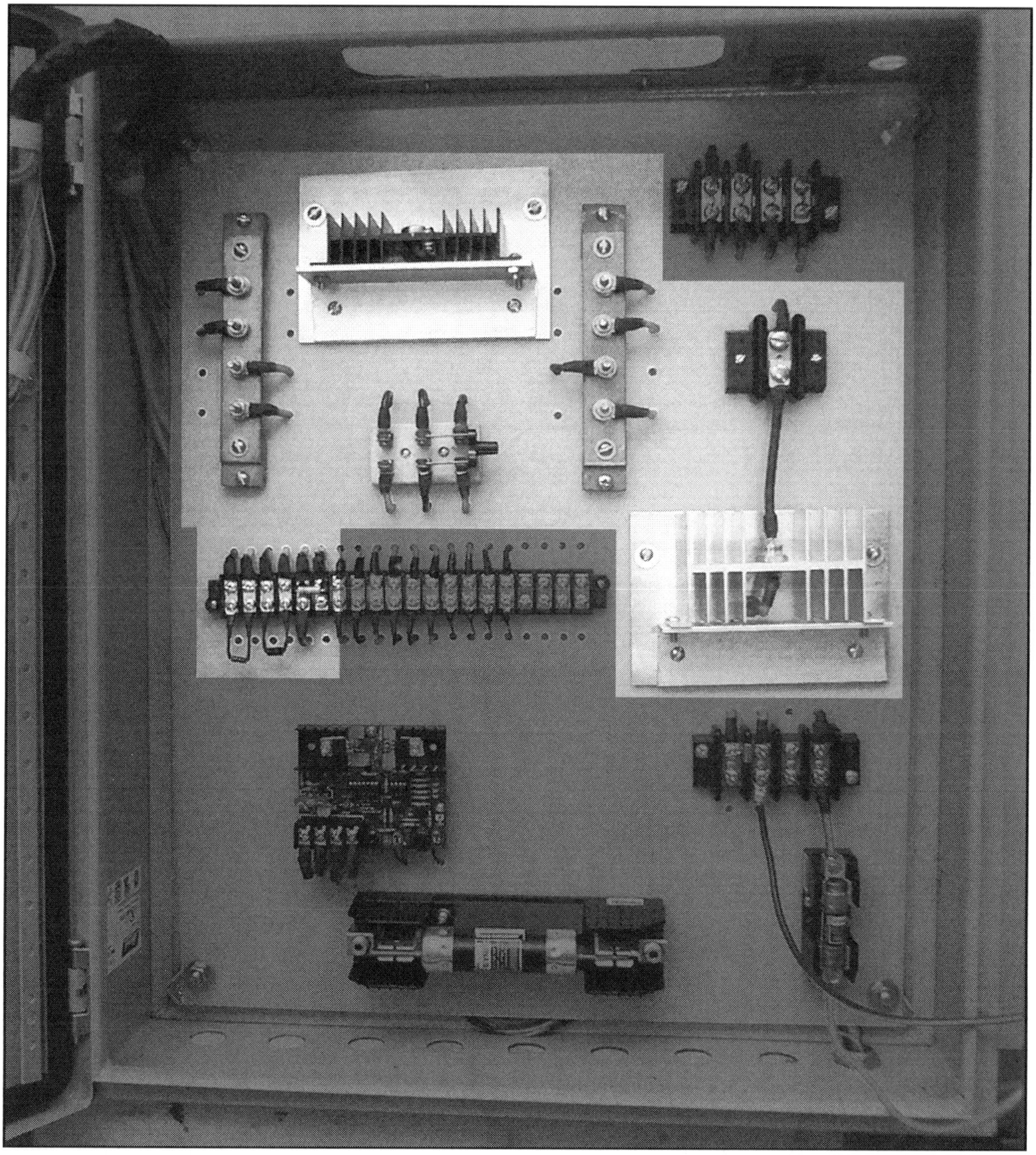

CC2 – single transistor switch shunt charge controller

CC2 consists of two bus bars, one resistor, a knife switch, two Zener diodes, one Schottky diode, and a Darlington transistor. This charge controller is suitable for flooded lead acid batteries, but should not be used with sealed lead acid batteries.

This circuit allows a regular charging cycle as well as manually controlled equalization. This is accomplished through the use of two Zener diodes, one rated at 14 volts and the other at 15 volts. Most Zeners have a tolerance of + 5%. This means that a 14 volt Zener can breakdown and begin conducting at anywhere from 14 volts to 14.7 volts. For instance, for the Zeners used in CC2, the breakdown was 14.5 volts for the 14 volt Zener, and 15.5 volts for the 15 volt Zener. 14.5 volts is a good general charging voltage for this type of regulator and 15.5 volts is a good equalization charge voltage.

The circuit operation is quite simple. When the battery reaches 14.5 volts, the Zener breaks down and allows current to flow to the base of the Darlington transistor, which then causes the current to flow through the collector and emitter of the transistor. This shorts out the panels and does not allow further charging to the battery.

When the voltage of the battery falls below 14.5 volts, the Zener does not conduct the current to trigger the transistor switch, thus the batteries can charge up to 14.5 volts again. The function of both Zeners is controlled by a knife switch. For normal charging operations, the current path will be through the 14 volt Zener. To equalize the batteries at the higher 15.5 volts, the knife switch is thrown to change the current path to route through the 15 volt Zener.

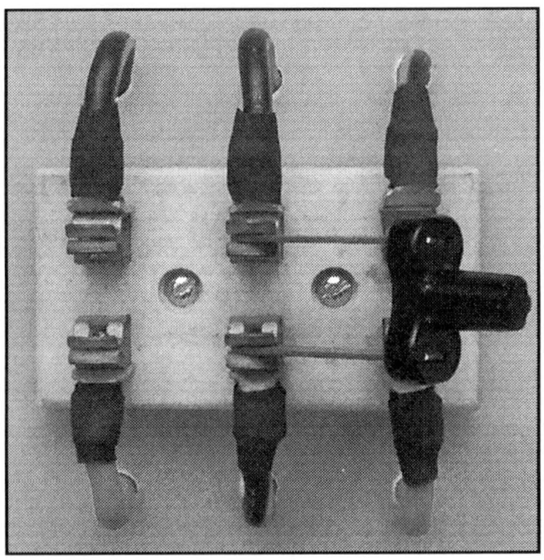

Knife switch, double throw, double pole

The Schottky diode in the circuit prevents reverse flow of current from the batteries back into the panels at night.

Heat sink assemblies

Heat sinks

The heat sinks were fabricated from aluminum plate, angle, and several small heat sinks that were on hand. For a system like this one, (using four PV panels as designed in *Build Your Own Solar Panels*), a heat-sink that is rated for about 300 watts for both the diode and the transistor is called for, and can be purchased from surplus suppliers (see page 180).

Surplus heat sinks come in a variety of shapes and sizes, and many have component holes for power transistors and diodes already drilled. Some of the holes for diodes may be smaller than needed, but they can be reamed or drilled through to enlarge the hole.

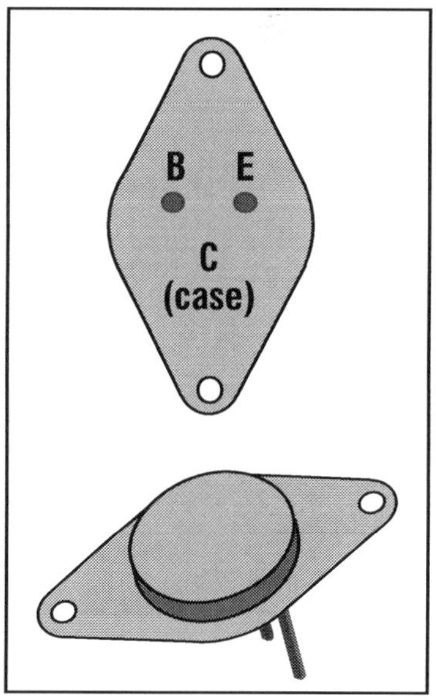

Darlington transistor

If you have a surplus heat sink for the power transistor and it does not have holes, the mica insulator from a transistor mounting kit can be used to draw the outline and template for the necessary drill holes. Some suppliers provide heat sinks with a power diode, see the list at the end of this chapter.

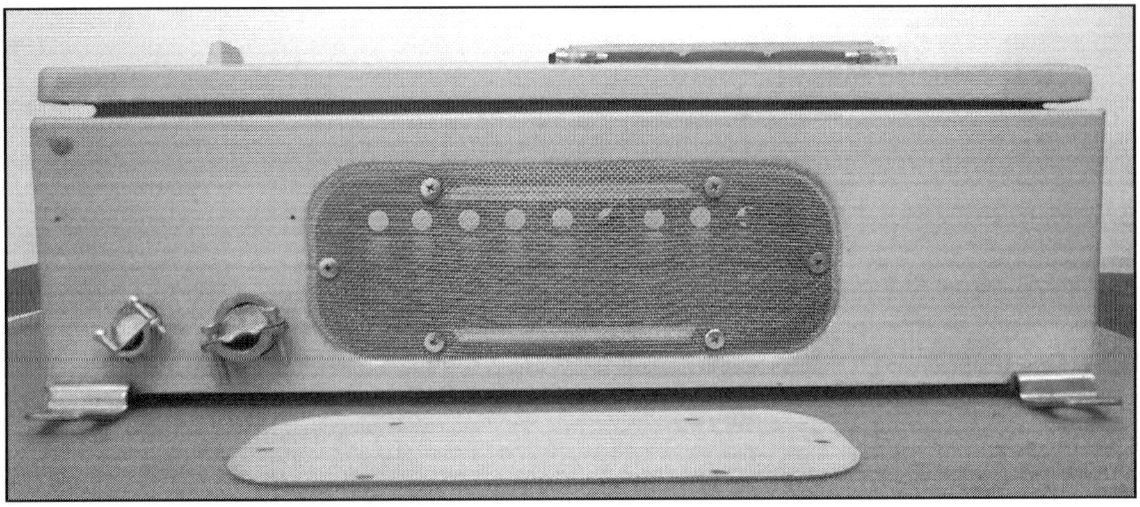

Screen vent in the top of the DC service/distribution box

Heat sinks and ventilation in the DC service box must be able to handle all the panels in the system's array. Our DC box has holes at the bottom and a screen vent in the top that create convective cooling. This works well, but very small muffin fan with or without temperature or other automatic control could be added to move the warm air out of the box and draw cooler air in. The muffin fan would run off the batteries.

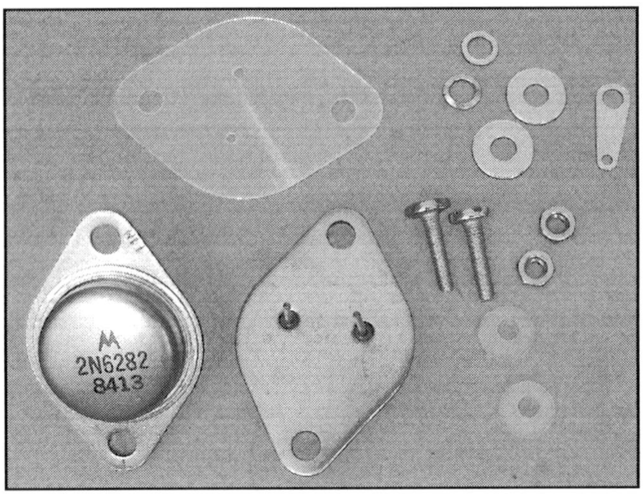

Darlington transistor kit

Insulating the diode and the transistor from its heat sink is optional. The difference is that not insulating the diode or transistor with a mica insulator will allow better heat dissipation, but will also make the heat sink live, so care would be needed

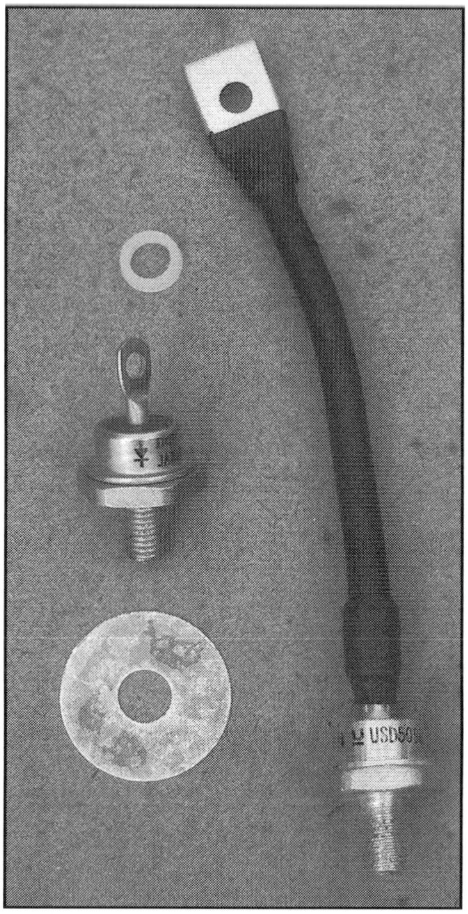

Diodes

not to short out the heat sink with tools to other components in the box. To avoid having a live sink, apply the mica insulator to these devices to ensure insulation.

The heat sink itself is isolated from the rest of the board as it is connected to the FR-4 which is non conductive. The attaching screws are not connected or contacting any metal parts or non insulated wires in the box.

For the transistor, the insulating bushings that come with a mounting kit can be used to help keep the pins of the transistor from touching the heat sink. Whether bushings are used or not, it is important to keep the base, emitter and collector electrically isolated from each other. In most cases, unless a crimp connection is used, you will be soldering wires to the transistor pins, and the stranded wire has to be wrapped around the pin before you solder. Make sure that no loose strands are connecting with other parts of the transistor or the heat sink. Insulation can be added to the connections by applying heat shrink tubing or liq-

Power transistor with heat sink

uid electrical tape around connections. Always apply a thermal joint compound between the diode or power transistor and the heat sink to ensure good heat conduction.

Schottky diode with heat sink

Charge controllers

Tools

Soldering station
Solder

Materials

CC1 charge controller kit available from CirKits (CirKits #CC3)

CC2 charge controller parts

ITEM	SOURCE	PART#	HOW MANY
ZN1 – Zener diode, 15 volt, 5 watts 1N5352 or ZN2 - Zener diode 14 volt, 5 watts 1N5351	Mouser Electronics or Circuit Specialists	1N5352 or 1N5351	1
Q1 – Darlington NPN transistor, 2N6282 or 2N6284	American Micro Semi Conductor for 2682, Circuit Specialists for 6284	2N6282 or 2N6284	1
D1 – Schottky diode, 75 amp, 45 prv	Surplus Sales Of Nebraska		1

CC2 charge controller parts, continued

ITEM	SOURCE	PART#	HOW MANY
R1 – 1000 Ohm, ½ watt resistor	Electronics store		
Heat sink for 2N6282 or 2N6284 rated for at least 300 watts	Surplus Sales Of Nebraska	HSK 350 watt or similar	1
Heat sink for Schottky diode rated for about 300 watts	Surplus Sales Of Nebraska (Also Fair Radio Sales)	Any heat sink	
TO-3 Mounting kit for 2N6282 or 2N6284	Electronics store		
Heat sink compound	Electronics store	10 8121	
TBC (terminal block C) – 1 position terminal block or 2 position terminal block if 1 position not available, such as 300 VAC/VDC terminal block 2 circuits, .69" center-to-center, 65 amps	McMaster-Carr		1
Steel electrical enclosure box	Surplus Center		
S3 – DPDT (double pole, double throw) knife switch, 25 amp rating.	Sargent Welch or Filnor or other	Sargent Welch	1
BB1, BB2 – 4 connection bus bars	See bus bar section or McMaster-Carr catalog		
Diode with heat sink, 30 amp, 150 volt rating (optional)			

Schematics

Solar II schematic legend

All of these circuits were integrated and routed via bus bars, terminal blocks, and switches as indicated in the text and schematics, and as shown in the photographs.

BB1 - Bus bar, 4 connections

BB2 - Bus bar, 4 connections.

ZD1 - Zener diode, 15 volt, 5 watts, 1N5352

ZD2 - Zener diode, 14 volt, 5 watts, 1N5351

D1 - Schottky diode, 75 amp/45 PRV

F1 - Fuse, 30 amp, class R and fuse holder

F2 - Fuse, 1 amp, and in-line fuse holder

F3 - Fuse, 1 amp, and in-line fuse holder

F4 - Fuse, 100 amp, class R and fuse holder

Q1 - Transistor, Darlington NPN, 2N6282, or 2N6284

R1 - Resistor, 1000 Ohms, 1/2 watt

S1 - Switch, single pole, double throw, (SPDT), toggle, on, off, on positions, 50 amp rating

S2 - Switch, single pole, double throw, (SPDT), toggle, on, off, on positions, 50 amp rating

S3 - Switch, double pole, double throw, (DPDT) knife, or other type. 25 amp or more rating

S4 - Fan switch

FN - Fan

PRG - Panel rack ground

LA - Lightning arrestor

INV - Inverter

AC - AC service box

CC1 - Charge controller (kit)

B - Battery/battery bank

M1 - Digital volt meter

SP - Solar panel/array

TH - Thermistor (CC1 kit)

LED - Light emitting diode (CC1 kit)

Heat sinks for Q1 and D1 are not shown in schematic

Terminal blocks A, B, C, D, E, are not shown in schematic at right

Complete Solar II System

- **BB1, BB2** - Bus bar
- **ZD1, ZD2** - Zener diode
- **D1** - Schottky diode
- **F1** - Fuse, 30 amp
- **F2, F3** - Fuse, 1 amp
- **F4** - Fuse, 100 amp
- **R1** - Resistor
- **S1, S2** - Switch, SPDT
- **S3** - Switch, DPDT
- **S4** - Fan switch
- **Q1** - Transistor, Darlington NPN
- **FN** - Fan
- **PRG** - Panel rack ground
- **LA** - Lightning arrestor
- **INV** - Inverter
- **AC** - AC service box
- **CC1** - Charge controller (kit)
- **B** - Battery/battery bank
- **M1** - Digital volt meter
- **SP** - Solar panel/array
- **TH**- Thermistor (CC1 kit)
- **LED** - Light emitting diode (CC1 kit)

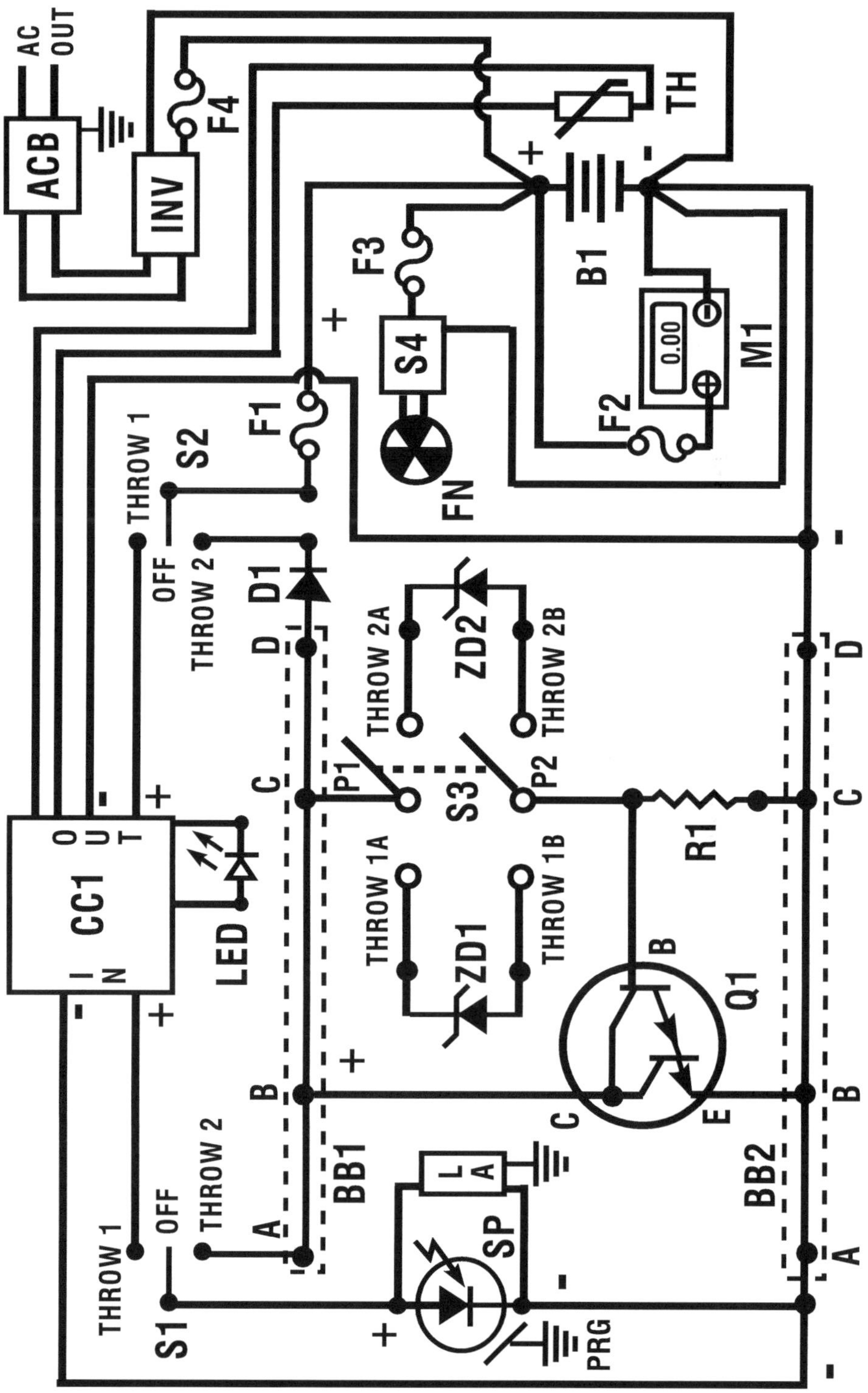

CC2 – Charge controller 2

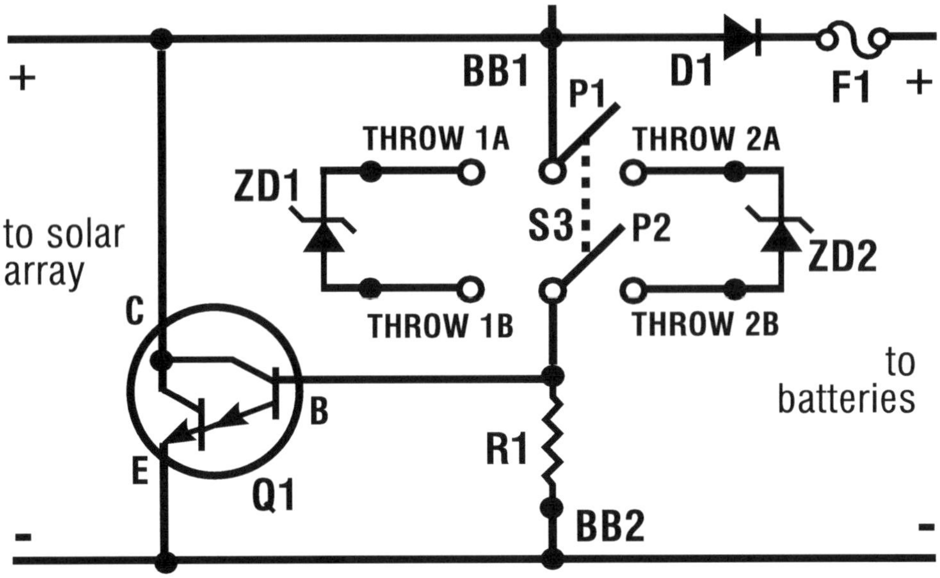

BB1, **BB2** – Bus bars
ZD1 – Zener diode, 15 volt, 5 watts, 1N5352
ZD2 – Zener diode, 14 volt, 5 watts, 1N5351
D1 – Schottky diode, 75 amp, 45 PRV
F1 – Fuse, 30 amp, class R
Q1 – Transistor, Darlington NPN, 2N6282, or 2N6284
R1 – Resistor, 1000 ohms, ½ watt
S3 – Switch, double pole, double throw, (DPDT) knife, or other type, 25 amp or more rating

BB1 – Bus bar 1

POSITION	CONNECTS TO
A	TBA, position 8
B	collector on Darlington transistor
C	pole 1 of DPDT knife switch 3
D	TBC, position 1

BB2 – Bus bar 2

POSITION	CONNECTS TO
A	TBA, position 2
B	emitter on Darlington transistor
C	TBC, position 11
D	TBD, position 3

TBA – Terminal block A See page 112

TBB – Terminal block B

POSITION	CONNECTS TO
1	throw 1B
2	anode ZD1
3	throw 1A
4	cathode ZD1
5	throw 2B
6	anode ZD2
7	throw 2A
8	cathode ZD2
9	Darlington transistor collector, also to R1
10	P2
11	BB2C
12	no connection
13	fan switch, positive in (SW 4)
14	battery positive
15	fan switch, negative in (SW 4)
16	battery negative
17	fan switch positive out (SW 4)
18	fan positive
19	fan switch negative out (SW 4)

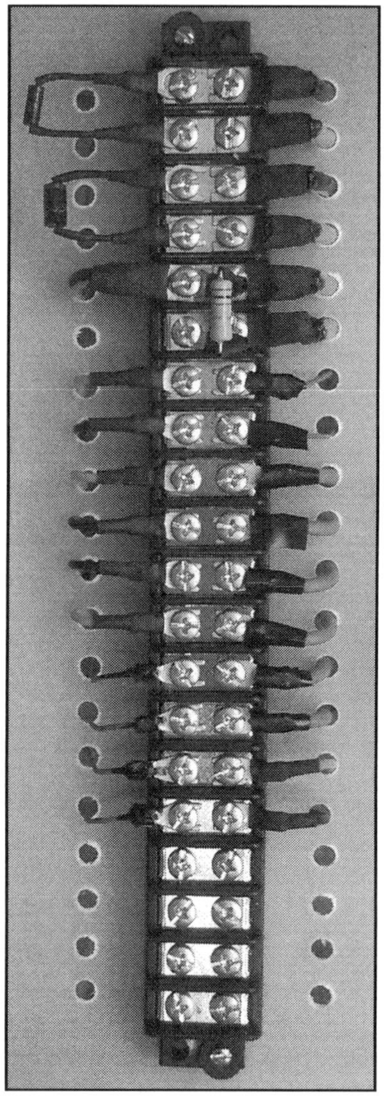

Terminal block B

TBB continued on next page

TBB continued

POSITION	CONNECTS TO
20	fan negative
21	LCD voltmeter positive
22	battery positive
23	LCD voltmeter negative
24	battery negative
25	LED positive lead
26	LED positive on CC1 printed circuit board
27	LED negative lead
28	LED negative on CC1 printed circuit board
29	temperature sensor port on CC1 printed circuit board
30	temperature sensor at battery
31	temperature sensor port on CC1 printed circuit board
32	temperature sensor at battery

Terminal block B

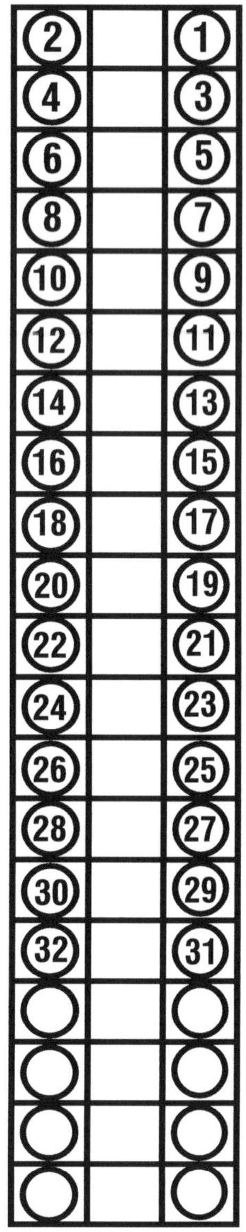

TBC – Terminal block C

POSITION	CONNECTS TO
1	BB1D
2	Anode D1

Terminal block C

D1 cathode – connects to SW2 throw 2

SW2 throw 1 – connects to CC1 positive out

TBD – Terminal block D

POSITION	CONNECTS TO
1	CC1 negative output to battery, connected via jumper to position 3
2	no connection
3	BB2D, connected via jumper to position 1
4	B1 battery negative
5	no connection
6	no connection
7	pole of SPDT toggle switch SW2
8	F1 fuse 1

Terminal block D

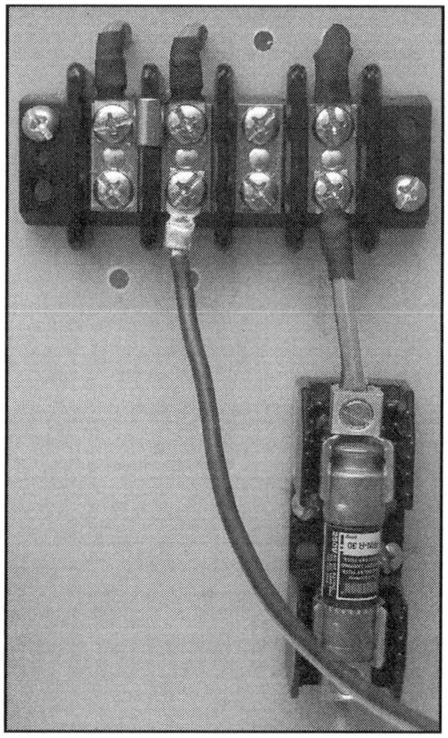

Terminal block D

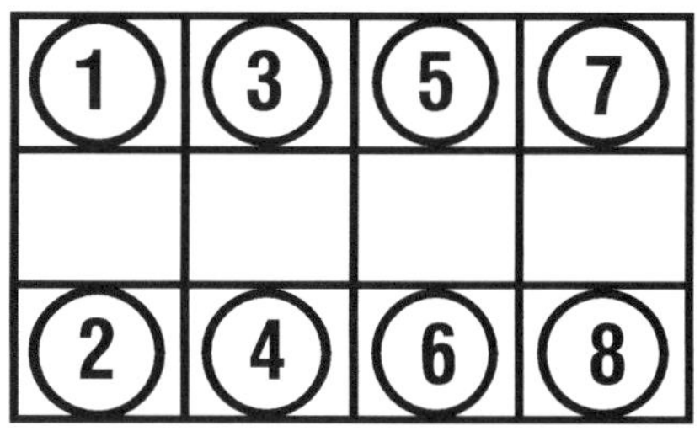

TBE – Terminal Block E in solar array combiner box, see page 99

Batteries

Batteries are electrochemical devices that store energy. Primary batteries, are non rechargeable, such as the common alkaline battery. Secondary batteries are rechargeable, such as the lead acid battery used in automobiles, or the nickel cadmium, nickel metal hydride, and lithium ion used in small electronic devices such as laptops and cell phones. Another type of rechargeable cell or battery, although not often used, is nickel iron.

PV systems that require storage of the energy they produce use secondary rechargeable batteries. Other methods of energy storage include super capacitors, hydrogen, and flywheels.

Although other types of rechargeable batteries are used in PV power systems, the most common is the lead acid battery.

Types of lead acid batteries

There are basically three types of lead acid batteries: SLI, marine, and deep cycle.

SLI (starting, lighting, ignition) is the commonly available type of lead acid battery used in cars and trucks. This type of battery has thin plates and is designed for applications where the battery will not be discharged below 20% of its rated capacity. This type of battery is not a good option for PV systems unless it will provide only sporadic power. They really should not be used unless there is no other alternative, or if power requirements are extremely minimal.

Marine batteries are middle of the road between SLI and deep cycle in that their plates are thicker than SLI plates, but they are thinner than deep cycle battery plates, so they are more suitable for PV systems than SLI batteries, but still not adequate for a serious power system.

photo courtesy Independent Power & Light

Deep cycle battery

Deep cycle batteries are what should be used for a serious PV power system. They have thicker plates and are designed for deep discharge of 50% to 80% without damage to the battery. Deep cycle batteries are commonly used in golf carts, fork lifts as well as alternative power applications.

In some parts of the world it is hard to get marine or deep cycle batteries, so there may be no choice but to use SLI batteries, which are more readily available. In more developed areas, marine batteries are easily available at a reasonable cost. Most large outlet stores carry them.

In the long run, it is more costly to use SLI and marine batteries for PV systems. They do not last as long for power applications and have to be replaced much more often than true deep cycle batteries.

Sealed or flooded deep cycle batteries

Deep cycle batteries are either sealed, such as gel cells or AGM (absorbed glass matt); or flooded. Most PV power systems use flooded lead acid batteries because its easy to ascertain the state of charge with a hydrometer, and they are not as prone to damage from overcharging as the sealed types.

With sealed batteries, charging must be precise and done carefully. On the other hand, they require less maintenance as electrolyte levels do not have to be monitored and distilled water doesn't have to be added periodically.

It is a matter of preference whether to use sealed or flooded types. I prefer flooded types as they are more forgiving and I do not mind checking fluid levels regularly. However, if you want to avoid this task you will prefer a sealed battery.

Another reason I prefer flooded lead acid is that I do a lot of experimentation and I do not have to worry about frying the batteries with an occasional mistake. If you are putting together a system for the first time and building a charge controller, I recommend using flooded lead acid batteries.

Purchasing batteries

Batteries are heavy and thus expensive to ship, so buying locally is usually the best deal. You can find dealers in your area by contacting the manufacturers and inquiring for the nearest point of purchase for you. Another option is to purchase batteries from a local PV installer. They usually sell deep cycle batteries and can also be good sources for other system components you might need.

There are several very good brand name deep cycle battery manufacturers. The most well known is Trojan. They have been providing quality deep cycle batteries at a relatively reasonable price to the alternative energy market for quite a few years, and have a proven track record with their product. There are higher quality batteries available, but they are also extremely expensive. The Trojan line allows the average consumer a reasonable startup cost, and they have a wide distribution network.

Configuring a battery array

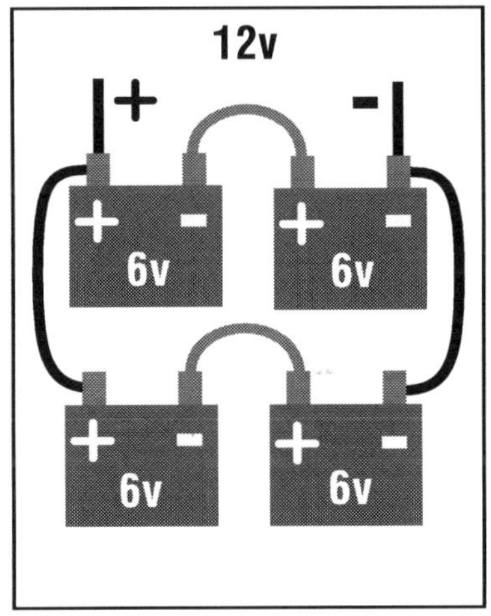

Battery array configuration for Solar II

For the Solar II project, the battery array consists of four Trojan T-105 6 volt 225 amp hour batteries. They were configured as a 12 volt system by making two series connected pairs, with the pairs connected to each other in parallel to provide 12 volt, 450 amp hours.

Batteries of other voltages can be used to make an array, such as 12 volt, or individual industrial type cells that are rated at 2 volts each. You can also work with a variety of smaller lead acid 2, 6 or 12 volt cells or batteries. Individual cells and batteries come in a variety of shapes, sizes and output ratings, and can be fun to experiment with for different applications.

When setting up a new system, always try to use new batteries. Older batteries may not perform very well, so it's really taking a chance if you do not know how they were used and if they were maintained properly. But, if the price is right and you do not require sterling performance it might be worthwhile.

Also, change all batteries in the array at the same time. If the batteries in an array perform at different levels, the worst battery in the array will bring down the others. The best practice is to start with new batteries and replace the whole array when needed.

Lead acid, deep cycle battery basics

Most lead acid batteries are composed of more than one cell. Each cell is rated at about 2 volts. A 6 volt battery will contain three cells, a 12 volt battery will contain 6 cells. These cells are connected in series to make the final output voltage.

Each cell is composed of two electrodes isolated from each other by a separator, in a container with an electrolyte. The negative electrode is composed of sponge lead, and the positive electrode of a fully charged battery is composed of lead dioxide.

Electrolyte in lead acid batteries

Lead acid batteries use an electrolyte composed of distilled water with about 28% to 36% sulfuric acid. During the charge and discharge cycle of a battery the percentage of acid to water varies in the electrolyte. When the battery is fully charged, the electrolyte contains the highest percentage of sulfuric acid. When the battery is near full discharge the electrolyte consists mainly of distilled water. As the battery discharges, lead sulphate is formed on the plates of the battery. When the battery is recharged, the process is reversed and the lead sulphate diminishes as the water in the electrolyte becomes more acid.

This change in sulfuric acid and water concentration in the electrolyte is an indicator of the state of charge of the battery, so the state of charge can be measured with a hydrometer, which reads the specific gravity of the electrolyte. Specific gravity is the measurement, in this case, of the ratio of the weight of the electrolyte to the weight of an equal volume of water at a specified temperature.

Sulphation

If a battery is not sufficiently charged during a charging cycle, lead sulphate remains on the surface of the plates. Repeated insufficient charging will cause lead sulphate to build up on the battery plates, a condition known as sulphation, which reduces the active area of the plates that can react with the electrolyte. This diminishes the capacity of the cell and thus the battery.

Also, all cells and batteries self discharge over time. Leaving a battery without charging it occasionally will result in a diminished level of charge and is another cause of sulphation on battery plates.

To protect batteries from sulphation, they should be given an equalization charge at least once a month. An equalization charge is charging the batteries at a higher than normal charge value. Usually a battery is charged at about 14.5 volts. An equalization charge is usually about 15.5 volts. This helps to bring the electrolyte concentration in each cell to the same level. In a series connected circuit, such as cells connected in a battery, the cell that produces the least energy will reduce the output of the other cells to its own level. So, it is important to equalize the output of each cell. Another option is to use an electronic desulphator on the batteries to prolong the life of the batteries.

All batteries sooner or later give up the ghost and will diminish in capacity. It is important to have some idea of what the average lifetime of a given battery is under standard use conditions. Most deep cycle batteries will give their best for about four to five years. After that it's touch and go. A battery replacement schedule should be figured into any estimate of system costs.

Charge cycles and battery life

For the Solar II project we used Trojan T-105 6 volt flooded lead acid batteries. One of the factors that determines the lifetime of this particular battery is the daily depth of discharge. The deeper a battery is discharged on a daily basis, the shorter the lifetime of the battery. For instance, with the T-105s, a daily depth of discharge of only 20% will give about 2,918 cycles. A daily depth of discharge of 30% will give 2,028 cycles. A daily depth of discharge of 50% will give 1,302 cycles, and a daily depth of discharge of 70% will give 973 cycles.

If you equate one cycle – one charge and discharge cycle – with one solar day, 1095 cycles equal three years, 1460 cycles equals four years, 1825 cycles equals five years and 2190 cycles equals six years and so on.

With this particular battery the minimum state of charge should not be below 30%. This means the batteries can be discharged to a depth of 70%. This of course is not advisable on a regular basis as it will greatly reduce the batteries' life span. If possible, try to keep the discharge depth at about 30% on a daily basis. This is an ideal, and in fact it will probably be more like 50%, which is acceptable as normal battery use. At the very least, with reasonable discharge depths you should be able to get about four to six years of life from a T-105 battery.

The float charge life of a T-105 is about 10 years. This means that if the battery is kept fully charged and rarely used, it would last about 10 years. If the battery is used at about 20% of its capacity it should last about 8 years. For certain low power, backup, and communications applications this would be a reasonable use.

Battery maintenance

Routine maintenance should include periodic water level and state of charge checks, equalization charging, checking contacts for corrosion and tightness, and keeping the batteries clean.

Safety and batteries

When working with batteries **always** wear proper protection including eye goggles and rubber gloves, and old clothes. A little bit of electrolyte always splashes when the vent caps are taken off. Expect that there will be inadvertent splashes of battery acid, and believe me, you do not want it contacting your eyes or skin.

An often overlooked hazard of working with batteries is the task of moving them. Batteries are very heavy and the weight is concentrated in a small area. Do not risk injury by carelessly trying to carry batteries around. Use a hand truck or some device that reduces or eliminates lifting. If you do carry batteries, use carrier straps that clip onto the battery, and have another person lifting with you.

Safe maintenance practice

Always use an insulated wrench to tighten and remove battery nuts. It is very easy to short out terminals and the current can quickly melt a wrench right before your eyes.

Do not smoke while servicing a battery and keep all sources of arcs, sparks and ignition such as open flames, heaters away from the

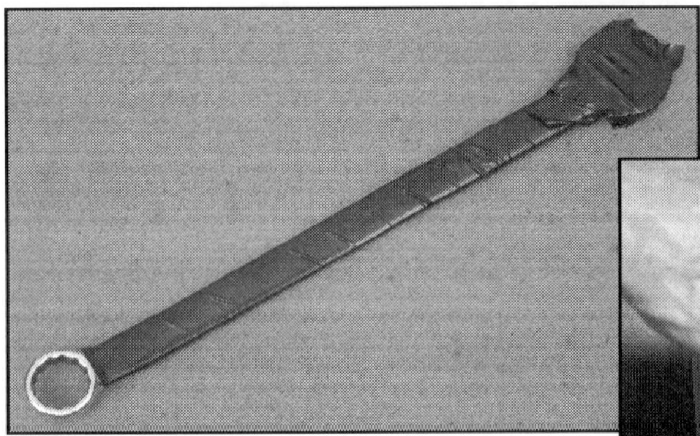

Always use an insulated wrench. A metal wrench can be covered with electrical tape.

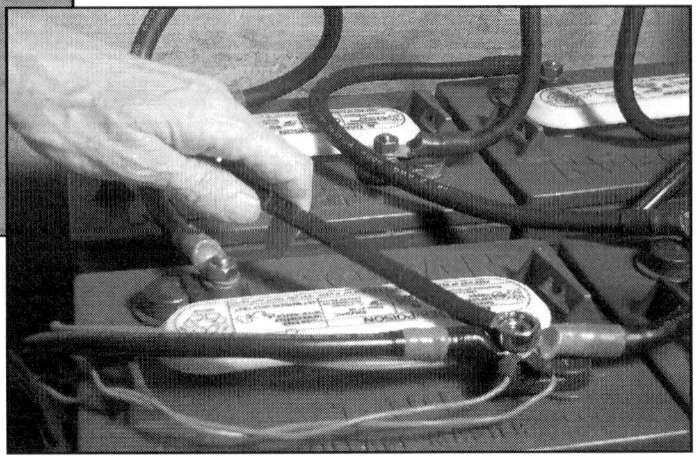

battery bank. The hydrogen and oxygen liberated from the battery water can be quite explosive.

Battery maintenance kit

A simple battery maintenance kit can be stored neatly next to the battery bank. It should include the following:

- Safety goggles
- Rubber gloves
- Hard toothbrush
- Post cleaner
- Multimeter
- Hydrometer with thermometer and temperature compensation table
- Insulated wrench to fit terminal nuts
- Distilled water
- Water filler bulb, or automatic dispenser
- Baking soda
- Petroleum jelly
- Small rags or paper towels

The small rags are used to clean the surface of the batteries. Pour a little bit of distilled water on the rag, then wipe down the battery. Dispose of the rag after use as it will probably contain small amounts of electrolyte.

Cleaning battery terminals

Baking soda is used to clean corroded terminals. If you notice even the smallest amount of corrosion, remove it with baking soda dissolved in some distilled water and applied to the terminal with a toothbrush or rag. The baking soda will dissolve the corrosion easily.

The baking soda can neutralize the acid in the electrolyte, so it is important not to get any baking soda or baking soda solution into the electrolyte wells of the battery. Never clean the terminals with baking soda when the vent caps are off, because you do not want to splash any into the wells.

Be sure to clean with straight distilled water after cleaning with a baking soda solution. This is to remove the baking soda from the battery surface and terminals. If there is an electrolyte spill, you can use the baking soda to neutralize the acid and make wiping up the spill safer.

Corrosion problems can be kept to a minimum by applying anti corrosion rings around the terminals. These are felt-like rings impregnated with chemicals that reduce corrosion at the terminals. They are placed around the terminals before the connectors are put on.

Other battery terminal maintenance

Each terminal and connector can be coated with Vaseline, which helps to keep corrosion from forming. Do not put Vaseline on the terminals before the connectors are put on. Make all of the connections first, fasten and tighten them, and then cover them with Vaseline. If applied before the connectors are installed, the Vaseline will be a barrier to conduction. If the connectors are extremely corroded, cut the connectors on the wires and crimp on new connectors.

Always make sure that terminal nuts are hand tight and solid, but not overly tight.

Measuring the state of charge

However the batteries are used, the state of charge should be checked at intervals to determine if the batteries are in optimum condition for the circumstances. To measure the state of charge, a hydrometer can be used, or observe the voltage when the batteries are at rest between their charge and discharge cycles.

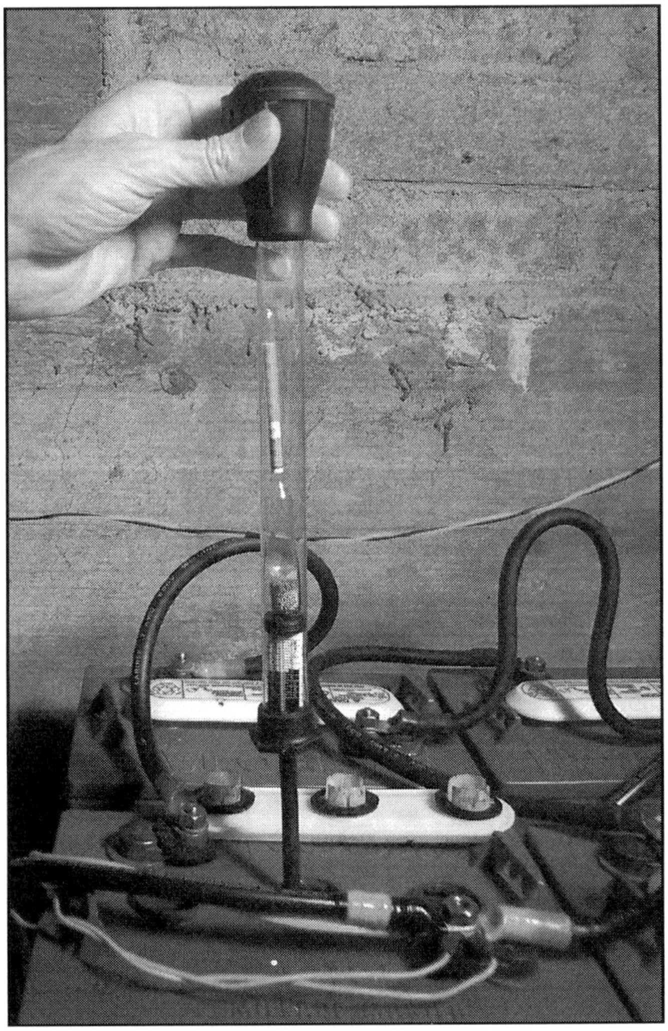

Using a hydrometer to measure specific gravity of the electrolyte

The hydrometer in a battery maintenance kit should have thermometer and temperature correction tables in it. When taking a reading, wait at least one hour from any discharge and six to twelve hours after the batteries have been fully charged.

When properly attired, remove the vent caps on the batteries and insert the hydrometer to withdraw a sample of electrolyte. Draw enough electrolyte so that the inner glass bulb in the hydrometer floats. The bulb will float higher or lower depending on the specific gravity. The reading is taken from where the surface of the liquid electrolyte meets the readout gauge.

Since we are working with T-105s, a reading is taken for each of the three cells in each battery. After taking the specific gravity reading and making the temperature correction for each cell, the value is written down, with a clear indication of which cell in which battery the reading is for.

The hydrometer readout is pretty straight forward. There are generally three colors (green on the bottom, white in the middle, and red on top) which indicate good, fair, and recharge. The finer gradations can be used with temperature compensation to get the most accurate reading.

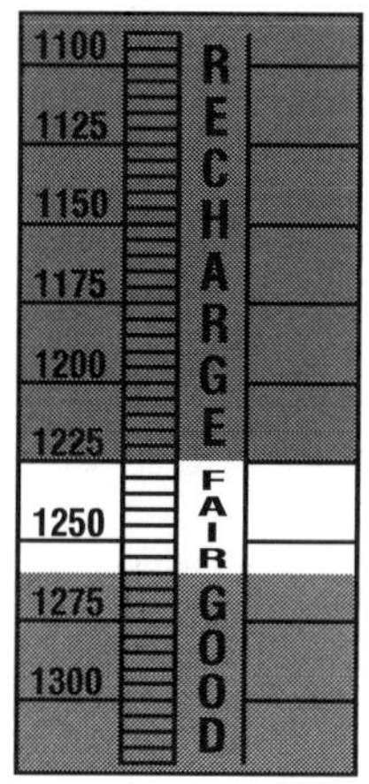

Hydrometer readout

Generally, at a temperature of about 75°F a battery that is 100% charged will read around 1.265. At 75% charge it will read about 1.239. At 50% charge it will read about 1.2. At 25% charge it will read about 1.17. These are just ballpark figures. Temperature compensation needs to be figured in for a correct analysis.

Another way to determine state of charge is to observe the voltage of the batteries after a period of no charge or discharge. Same as for the hydrometer method, wait about six hours before taking the voltage reading to measure the state of charge.

At a state of rest – a period of no charge or discharge for about six hours – and at a temperature of around 77°F, a battery that is charged 100% will have a voltage reading of about 12.7 volts. At 80% charge the voltage will be about 12.5 volts. At 60% charge the voltage will be about 12.2 volts. At 40% charge it will be about 11.9 volts and at 20% charge will be about 11.6 volts. Reading the state of charge by observing the voltage with the batteries at rest is not a precise measurement but can give you a good estimate.

Equalizing charge

Periodically, check the batteries' state of charge with a hydrometer as above to determine whether the specific gravity of the cells needs an equalizing charge. An equalizing charge is simply charging the battery to full capacity, and then overcharging at about 15.5 volts. This causes the batteries to gas a lot, which reduces sulphation and gets rid of stratification in the electrolyte.

Stratification is the separation of the water from the sulfuric acid in the battery cell. The sulfuric acid is denser and has a tendency to sit at the bottom of the cell, while the less dense water tends to stratify on the top of the sulfuric acid. During an equalization charge, the bubbling hydrogen and oxygen gasses that are created stir the acid and water together and reduce the stratification.

As noted earlier, sulphation is the excessive accretion of lead sulphate on the plates, which reduces the capabilities of the batteries. Some cells in the batteries will sulphate more than others. In order to equalize the output of each cell, the sulphate has to be removed from any cell that is not producing as well as the others.

The basic idea of equalization is to get the specific gravity of each cell to be the same as all the others in the battery, and to bring them all up to maximum charge and discharge capacity by removing as much lead sulphate from the plates as possible.

Opinions vary about how often to check the specific gravity of the cells with a hydrometer. When a system is first installed, the specific gravity should be checked each week for a couple of weeks, then a schedule of every 30 to 90 days should suffice. I suggest at least once a month, and at the same time, check the electrolyte level to see if more distilled water is needed.

If possible, measure the specific gravity of the cells after charging and no use for about six hours. After taking readings, note whether any cells differ from the others in readings and by how much they differ. If the readings vary .015 to .020, put the batteries on an equalizing charge for several hours. After the equalizing charge, check the specific gravity, and if there is still much of a difference from cell to cell, put them on equalizing charge again.

Adding water to the electrolyte

After equalization charging, check the electrolyte level and add distilled water if needed. When adding water, fill to about ⅛" below the bottom of the fill well. Do not overfill the cell wells, because when the batteries charge, overfilling can cause electrolyte to spew through the vent caps, which makes more corrosion and mess.

Never add distilled water to a discharged battery. When the battery is charged, the electrolyte level will rise and may cause leaks and spraying through the vent caps. Usually distilled water is

Adding distilled water to the batteries

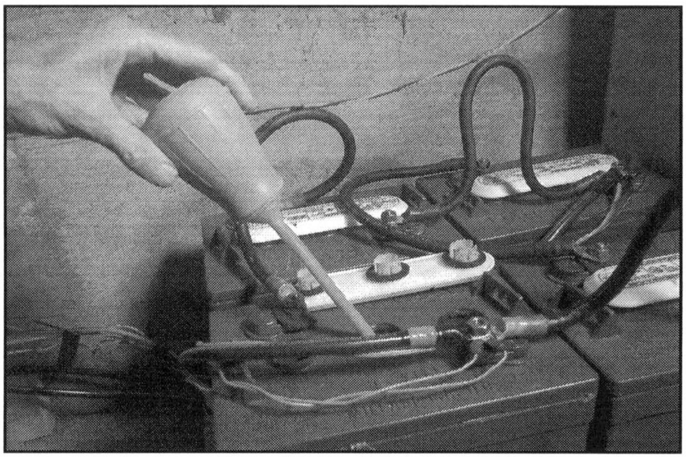

added after the equalization charge.

Make sure that the plates are covered in electrolyte, and that the cells' specific gravities are as equalized as possible to get the best charge and discharge performance from the batteries.

Every system is different and will require different time frames for equalizing depending on many different factors. During daily cycles, watch the voltage variations to discern patterns that will help you to understand how the batteries charge and discharge.

Rough guide for normal battery readings

BATTERY STATE	VOLTMETER READING
At or near full charge and charging	14.5
Somewhat discharged and charging	12.3 to 13.2
Light load on fully charged batteries	12.4 to 12.7
Fully charged with a heavy load	11.5 to 12.5

There is lots more information about batteries of different types on the internet and in catalogues from solar suppliers. If you're interested in details of maintaining, repairing, and building lead acid batteries for your PV system, I recommend *The Battery Builders Guide*, see page 183.

Tools for working with batteries

TOOL	SOURCE	PART #
Multimeter	Any electronics supplier, or hardware store	
Insulated wrench	Hardware store	
Hydrometer with thermometer/temperature compensation table	New England Solar Electric	721-004
Battery filler bulb	New England Solar Electric	721-005
(Optional) Battery filler dispenser	New England Solar Electric	721-007

Components for the battery array

ITEM	SOURCE	PART# AND QUANTITY
Battery interconnect cable	Backwoods Solar	O-BIC 2, 4 each
Battery anti-corrosion rings	New England Solar Electric	721-001 (for Trojan T105s), 8 each
Batteries - Trojan T-105, 6 volt, 225 amp hour	Backwoods Solar, other solar suppliers, local dealer	4 each

Battery Box and Vent

As batteries are being charged they release hydrogen and oxygen, and at some parts of the charging cycle, the gases are emitted in amounts that can be a safety issue if the batteries are not adequately ventilated. Although hydrogen dissipates upward rapidly, it is a good idea to enclose the battery array with a ventilated box. This prevents the gases from accumulating near the batteries, and also helps to remove corrosive electrolyte aerosols from the area. Most battery banks are located in a home, and it is not a good idea to be constantly breathing electrolyte fumes. With a small battery bank this may not be a problem, but that depends on the placement of the batteries in the dwelling, the air circulation, and other factors.

Batteries and ambient temperature

Battery banks should be located in an area that will not be exposed to weather and major temperature changes. Batteries suffer when placed in environments that are too cold or too hot. A mid range of about 77°F is ideal. It is generally recommended to keep them in a cool dry area. If batteries are in areas that get very cold, the battery box should be insulated. Also, note that at about 32°F hydrometer readings may not be very accurate.

Electrolyte can freeze in a battery, especially when the battery is in a discharged condition. In this state there is more water than acid in the electrolyte solution. Never try to charge a frozen battery. Warm it up before applying a charge. If batteries are going to be exposed to cold weather, be attentive and make sure that they are kept as fully charged as possible.

Battery box construction

Battery boxes can be made of any sturdy material, preferably non-conductive and acid resistant. The material most commonly used is wood. Acid resistant plastics can be used, but the materials are usually not locally available and are apt to be expensive.

Solar II battery box

An alternative is a combination of a wood structure with a lining of acid resistant plastic film, or coating the wood with an acid resistant compound.

The box for this project was made of plywood and 1½"x ¾" wood strips. The box is 24¾" long x 20" wide x 21" high. It holds four batteries with enough room for cable runs and space between the batteries.

The main part of the battery box was constructed with 5 pieces of plywood using wood strips to connect and frame the pieces. Screws were used to fasten each side to another via the wood strips.

There is a fan exhaust hole in the top, and two holes in the sides – one for wire and cable, and the other hole for air flow.

The bottom of the box is a separate piece. It has wood strips on the bottom side where it contacts the floor to provide some space so that water will not seep into the box in case of flooding or plumbing accident, since the box is located in a basement.

Wood strip stays were attached on two edges of the box bottom to keep the upper part of the box from moving side to side, and give it a snug and secure fit on the base.

The box was stained and coated with silicone on all surfaces. Urethane could also be used for a general coating. Wood battery box surfaces should be coated liberally and well to make a tight gas barrier. This will also protect the wood from acid aerosol, and from moisture. All of the joints inside the box were caulked with silicone to make it as gas and aerosol leakproof as possible.

The battery box, opened.

The air vent hole on the bottom side of the box should be at least 2" in diameter. The top exhaust hole can be cut in the center of the box top and should be cut so that a 2" PVC pipe fits in tightly. Two metal handles were secured to the top of the box for easy lifting.

photo courtesy Independent Power & Light

Typical battery box with removable lid

This design we used is unusual in that the bottom of the box is not attached to the rest of the frame. The upper part of the box is lifted off the base to service the batteries, so the very top of the box is sealed to the sides. Hydrogen gas from the batteries moves upward, so for a battery box to be effective in managing the hydrogen and acid aerosols, all the joints on top need to be very well sealed. Most boxes have a hinged or removable lid on the top, and/or one on the side. Although a hinged or removable lid is convenient, it can be almost impossible to make such a box hydrogen leak proof. To get a gas proof seal with a hinged lid takes exacting workmanship. Various types of foam sealants, sealant strips and tapes can be used, but they wear with time and have to be replaced at periodic intervals.

Tools for building the battery box

- Saw
- Caulking gun
- Metal cutters
- Hole saw for entry cut
- Drill
- Screwdriver

Battery Box and Vent

Materials for the battery box

- Silicone caulk
- Wood stain (optional)
- Urethane (optional)
- Plywood
- Wood strips
- Handles
- Screws

Ventilation system

The exhaust system connected to the battery box consists of PVC pipe and fittings, and a fan unit.

PVC pipe parts:

- 1 – 45° elbow
- 1 – 90° long elbow
- 2 – pipe couplers
- 1 – length of flexible PVC pipe
- 2 – cuffs for ends of flexible PVC pipe
- 3 – connectors, several inches long, cut from 2" pipe
- 1 – retaining ring cut from 2" pipe to hold screen in 45° elbow
- 1 – 2" pipe feed-through, length depending on need

Also:

- 2 – round screens, cut from metal window screen stock
- 1 – fan unit

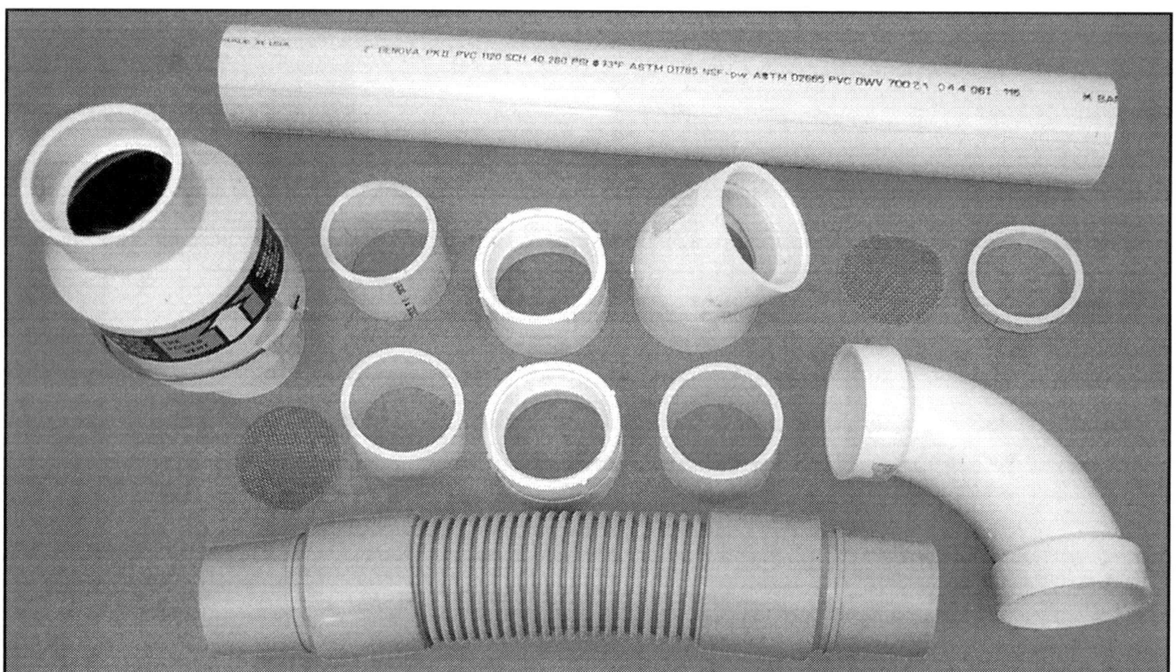

Parts for the vent

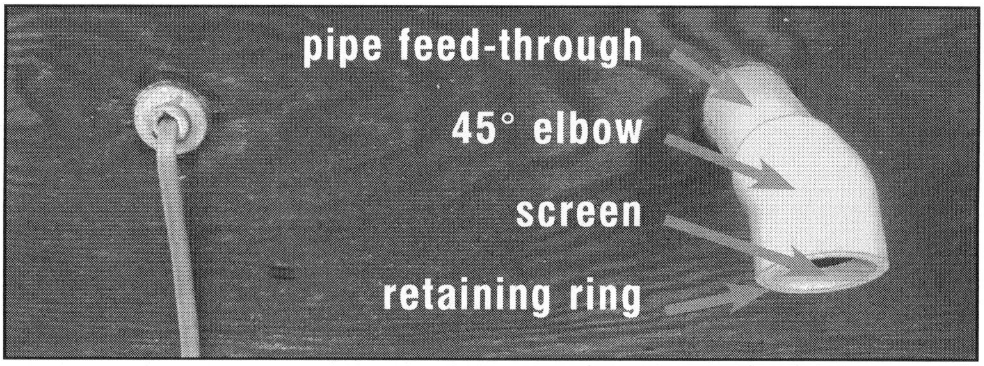

Right, outside of battery vent; UFB entry at left

The outside vent of the exhaust system

Screen and retainer ring

1. A piece of 2" PVC pipe was inserted into the 45° elbow and seated against the inside lip of the elbow.
2. The outside of the 2" PVC pipe was marked where they join.
3. The pipe was removed from the elbow and the pipe was cut just inside the mark. This small piece is the retaining ring to hold the outside protective screen against the lip of the elbow.
4. The retaining ring was laid on a piece of metal window screen and the outer circumference outlined on the screen with a marker.
5. Metal sheers were used to cut along the outline.
6. The screen was fitted by inserting it into the elbow and seating it easily but firmly on the inside lip of the elbow.
7. The outside of the retaining ring was coated with silicone and slipped into the pipe against the screen and the lip of the 45° elbow.

Feed-through

8. 2" PVC pipe was cut to length to go through the building wall, and extend on either side.

9. Wall holes were cut for the feed-through. Wall holes should be tight to the pipe and have as little play as possible.
10. The outside of the feed-through pipe was coated with silicone and inserted through the wall. The joint between the wall and the PVC pipe was coated with silicone both inside and outside the building to form a weather barrier and hold it in place. This was allowed to dry for 24 to 48 hours.
11. The outside vent elbow was fitted over the feed-through pipe with the opening facing downward. This prevents precipitation from entering the vent. This joint can be either siliconed or press fitted.

Indoor vent assembly

The fan unit case is made of PVC and was designed to accept 2" PVC pipe on each end. The flexible PVC pipe makes it possible to remove the 90° elbow from the feed-through pipe when the box needs to be lifted and moved to the side to service the battery bank.

Inside the fan vent

12. On the inside of the building wall, the long 90° elbow was press fitted to the feed-through, and a short piece of 2" PVC pipe was inserted into it, then the other end of this pipe was press fitted into a coupler.
13. To secure the fan to the battery box, a short piece of 2" PVC pipe (connector) was inserted into the vent hole in the top of the box. The fit was tight, and the edges of the

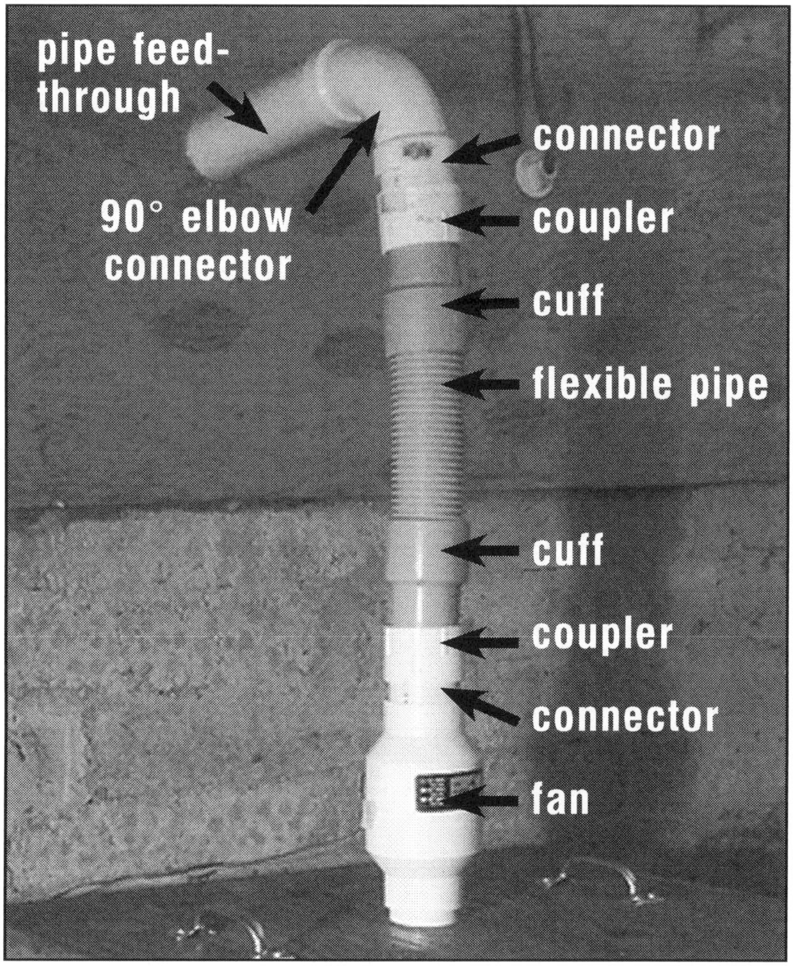

pipe were epoxied onto the box, and allowed to dry for two days. Inside the box, the end of the pipe was flush with the wood interior of the box top. It is important that such a vent not extend beyond the surface of the box interior. This is so that escaping hydrogen gas will have a smooth path out of the box into the pipe, and not build up in any corner at the top of the box.

14. The fan unit was fitted onto the connector PVC pipe on the box.
15. Another short piece of PVC pipe was inserted into the top of the fan unit, and a coupler was press fitted on top of that pipe.

16. Next, a cuff was placed on one end of the flexible PVC pipe and inserted into one of the couplers. The flexible pipe was cut to fit and a cuff fitted on the cut end.
17. The cuffs were fitted snugly into the couplers.
18. The switch for the exhaust fan was mounted in the DC service box.

Battery vent switch

This particular switch has an automatic and a manual on/off switch. In automatic mode, the fan will turn on as the battery voltage rises beyond 13.8 volts, and then turn off after about a 30 minute delay when the battery voltage goes down past 13.8 volts.

Automatic is the preferred mode of use since the actual state of the batteries controls the fan – it turns the fan on when the batteries start giving off more gas during charging, and shuts off when the gassing is reduced.

Tools for assembling and installing the battery vent

- Saw
- Caulking gun
- Metal cutters
- Hole saw for entry cut
- General wiring tools
- Materials for the battery vent

Materials for the battery vent

ITEM	SOURCE	PART# / QUANTITY
Metal window screen	Hardware store	
Length of 2" PVC pipe	Hardware store	
PVC couplers for 2" PVC pipe	Hardware store	2 each
Low temperature semi-flexible EVA vacuum hose 2" ID, 2⅜" OD, length depending on need	McMaster Carr	5397K131
Screw-on fitting for 2" ID low temperature semi-flexible EVA vacuum hose (cuff)	McMaster Carr	5397K19
45° PVC elbow	Hardware store	1 each
90° long elbow	Hardware store	1 each
Silicone caulk	Hardware store	
Exhaust fan	Backwoods Solar	O-PVENT 12
Exhaust fan control switch	Backwoods Solar	O-BATCONT12
Fuse 1 amp with in line fuse holder	Electronics store	
Inline fuse holder	Electronics store	

Inverters

There are basically three types of inverters available. True sine wave, modified square wave (sometimes incorrectly called modified sine wave) and square wave. True sine wave is the most expensive and produces AC power that is similar to what the power company provides. Modified square wave is close to what the power company provides but is not quite the same. Square wave is not anywhere near what the power company provides – but will suffice for limited applications with some AC appliances.

A true sine wave inverter

All AC appliances run very well on true sine wave inverters.

Most AC appliances run well on modified square wave inverters except for some stereo systems, some Mac computers, inexpensive rechargeable tools, light dimmers, devices with motor controls for speed, and laser printers.

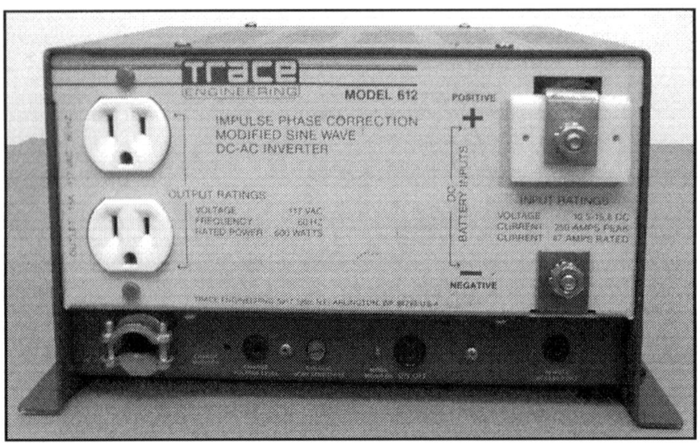

Modified square wave inverter

Modified square wave inverters can also cause interference with radio reception and produce distortion on some TVs.

Square wave inverters are best only used on resistive loads such as incandescent lighting and heating elements. They do well with some types of motors such as universal, brushless motors, but do not work with induction motors.

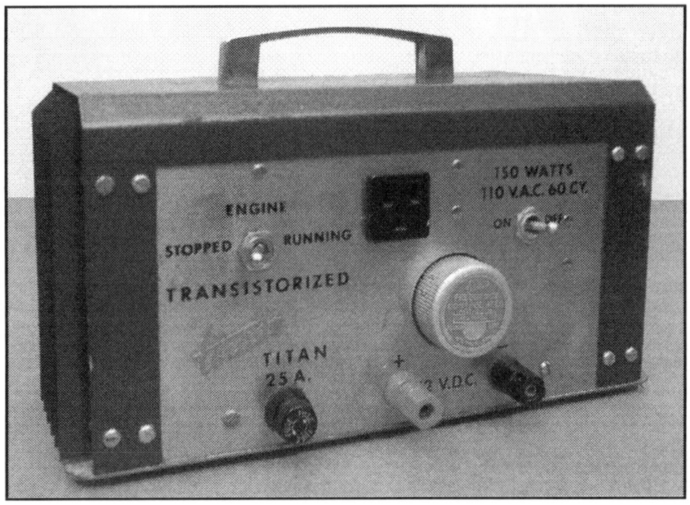

Square wave inverter

Most users of renewable energy choose modified square wave inverters for reasons of cost. They are much less expensive than true sine wave and for the most part work quite well with most electronics, computers, and electrical equipment. If you have a device that you prefer to run off a true sine wave, then you can buy a small true sine wave inverter to be used with that appliance only and rely on a larger capacity modified square wave inverter for everything else. This greatly reduces the initial cash outlay particularly when a true sine wave is not needed for anything but one or two small items.

Sizing an inverter

Large motorized appliances can draw up to twice their normal running power on startup. An inverter needs to have a surge capacity that can handle the start-up of any motors plugged in to it.

Installing an inverter

Modified sine wave inverters have come down in price in recent years and are currently very inexpensive, so it does not make too much sense to build one yourself. They do require some skill to fabricate, and you would probably not save much money considering the time and effort involved.

For the Solar II project we used an old 612 modified square wave Trace inverter that had been given to us. It had three things going for it:

- It was free.
- It was a 600 watt inverter, which is just the size needed.
- It was in excellent condition.

The Trace inverter used for Solar II, installed

By the way, it is good practice to make friends with people with solar systems. When they upgrade they often have equipment which is in very good condition that they'll sell at a more than reasonable price, or give away as they move on to bigger and better components.

This particular inverter had a hard-wire terminal strip for connections as well as two plug in receptacles for AC power. We used the hard-wire option as part of a permanent installation.

Each brand of inverter has different hookup and grounding configurations. They should be wired according to the manufacturers instructions.

For wiring from the inverter to the AC distribution box, and then to the outlets, 12 or 10 gauge NM, NMC, or UFB, can be used according to preference. We used NM. NM comes with a black wire, a white wire and a bare wire. The black wire is the "hot" wire. The white wire is the neutral or common wire and the bare wire is the ground wire. Each of these wires is connected to the appropriate terminals in the inverter. Our inverter has a terminal block connection for the hot wire, common/neutral outputs, and a screw and nut terminal for ground connected to the chassis.

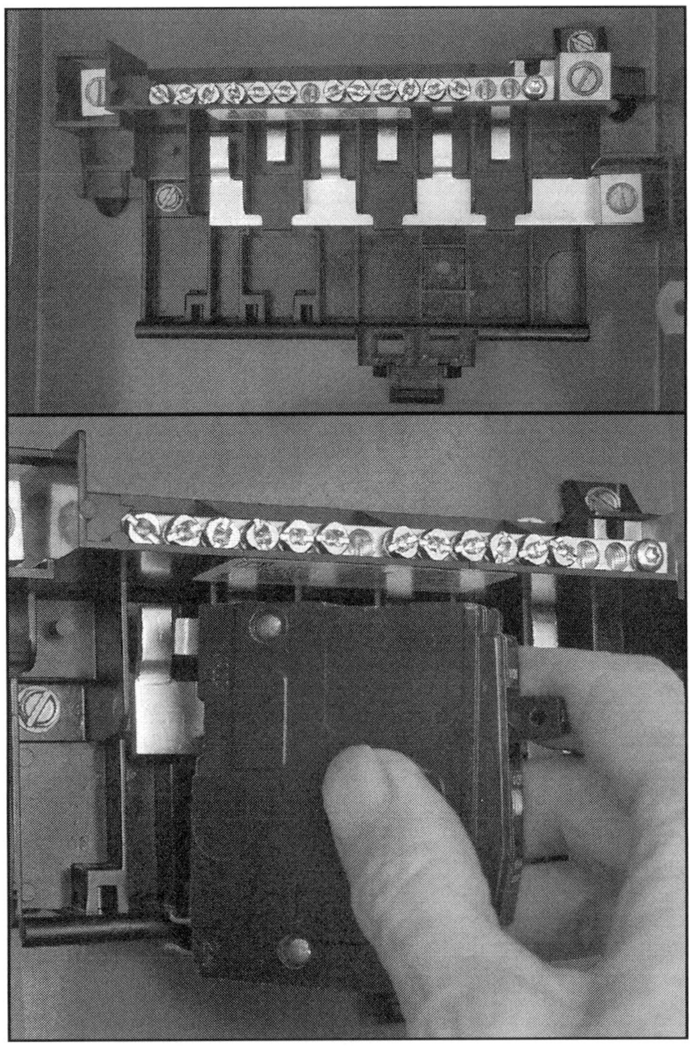

One AC breaker was installed in the load center of the AC service box

There are two screw terminals on the outside of the inverter case for input. They attach to the DC positive and negative cables from the battery bank. Two 2 gauge, 10 foot length

Inside the AC box with one breaker installed; left, without the cover plate; right, with cover plate in place.

pieces of inverter cable, with terminal lugs attached for easy connection, were attached to the inverter and battery bank. The length of the cable run from the battery bank to the inverter should be no more than 10 feet.

The positive cable was cut in half and stripped so that the cut ends could be placed into the 100 amp fuse holder in the DC service box. The positive cable from the battery was connected to the fuse in the DC service box and routed out of the DC service box to the inverter from the fuse. (All output connections from the positive battery terminal should be fused, and all the connections to the inverter and battery must be tight.)

AC service box

Next, the AC service box was wired to the inverter. The AC service box was a QO612L1100RB Square D circuit breaker load center. This load center has 6 spaces for breakers. The number of breakers you install will depend on the number of outlets you will be installing – each wire run from the AC service box to individual outlets should have its own breaker. This system would initially have only one outlet, so one breaker was installed. Installing breakers is simple. They are simply clipped into the spaces provided.

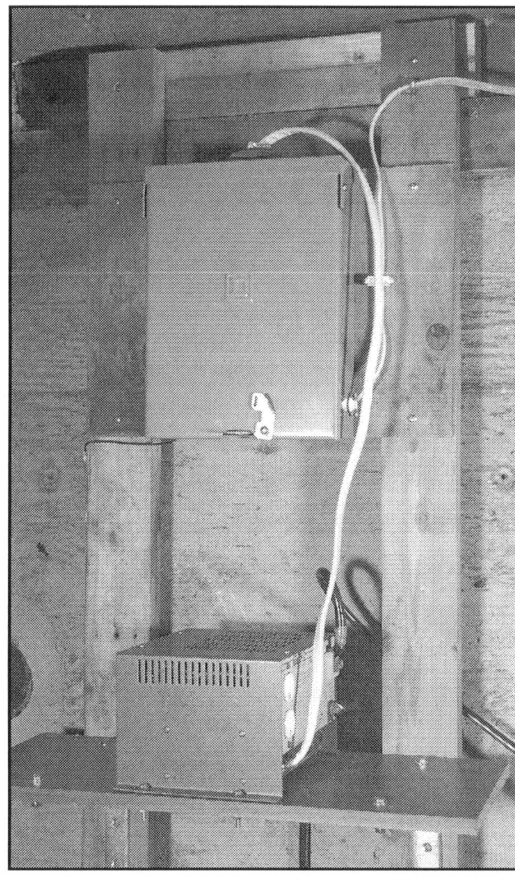

Inverter and AC distribution box

To connect the inverter to the AC distribution box, the hot (black wire) from the inverter was connected to the circuit breaker bus bar and the common and ground wire was connected to the ground/common bus bar in the box.

For the output to the outlet receptacles, the hot (black) wire was attached to its circuit breaker, and the ground (bare wire) and the white wire (common or neutral) to the ground/common bus bar. Then, a 6 gauge wire from the ground/common bus bar was attached to a good earth ground.

Next, the wire was run to the outlet and plug receptacles were installed. The details of this depend on your needs and the structure involved. Your local home supply store will have plenty of do-it yourself information, tools and options.

If you are going to have a DC run for DC appliances, you can use a QO612 distribution box that is rated for DC. However, do not mix DC and AC in the same box. Use separate boxes.

Tools for installing the inverter

Needle nose pliers

Parts for installing the inverter

ITEM	SOURCE	PART NUMBER
Inverter	Solar component suppliers	
Inverter cable	Backwoods Solar	BC10-2
AC distribution box DC distribution box	Backwoods Solar, McMaster Carr or hardware store	Backwoods Solar O QO612

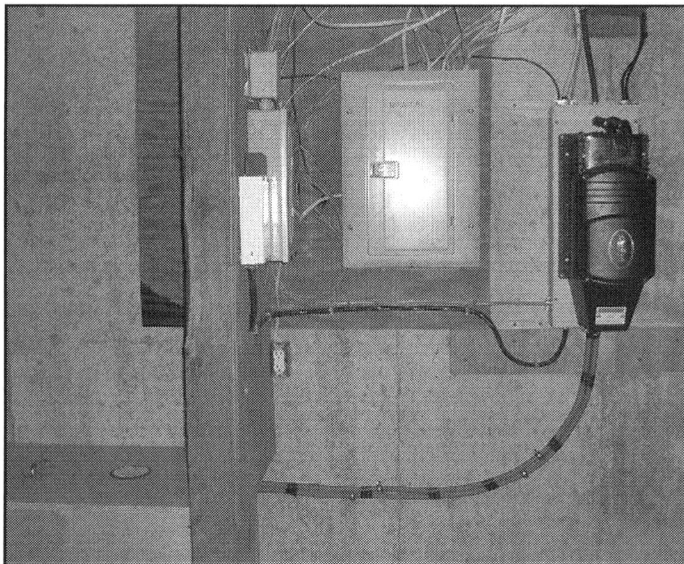

photos courtesy Independent Power & Light

Three different examples of power centers.

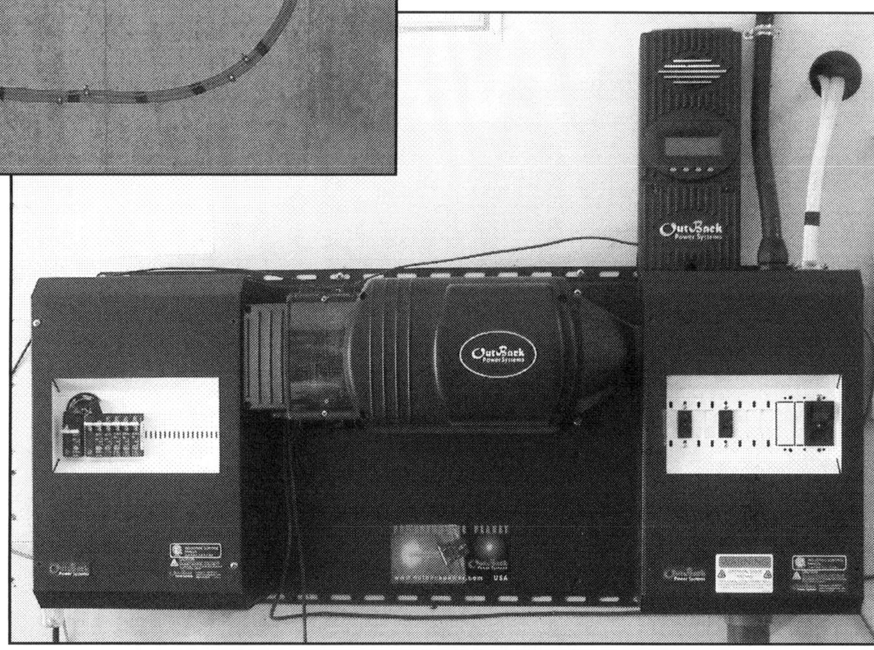

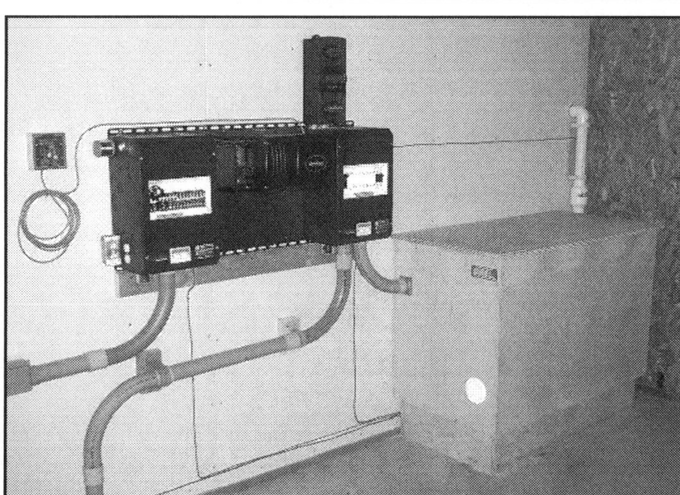

Appendix

Initial hookup sequence

1. Before connecting any wires from the panel input or battery in the system, make sure both toggle switches in the DC service box are in the off position.
2. Make sure the S3 knife switch is in the throw 1 position for normal charging regulated through the 14 volt zener.
3. Connect UFB positive to TBA3.
4. Connect UFB negative to TBA1.
5. Connect the two DC positive 5 foot inverter cables to each end of the fuse holder.
6. Wire inverter to AC distribution box, insert breaker, make sure it is in the on position.
7. Complete wire runs from AC distribution box to AC outlets.
8. Inspect all wiring and connections.
9. Check to see that inverter is turned off.
10. Hook up negative inverter cable, input wire from TBD4, fan negative wire, volt meter negative wire to negative terminal of battery bank.
11. Hook up positive inverter cable from fuse F4, input wire from fuse F1, fan positive wire, volt meter positive wire, to positive terminal of battery bank.
12. Hook up positive inverter cable from fuse F4 to inverter.
13. Label each switch and its throw position. Wire toggles so that when both switches are thrown in the same direction, they connect to the same charge controller.

Test sequence with two charge controllers in the system

1. Pick a sunny day
2. Note battery voltage reading on volt meter
3. Throw toggle switch S2 to throw 1 or 2 depending on which controller you are going to use.
4. Throw toggle switch S1 to throw 1 or 2 depending on which controller you are going to use.
5. If using CC1, the LED will light on the panel. Adjust the charge controller's potentiometer to align the float voltage set point.
6. If using CC1 there is no charge indicator light, so readings will have to be taken with a volt meter which should rise indicating charging.
7. On a good sunny day, the volt meter will rise to the preset voltage, around 14.5 volts, and maintain at that level for normal charging. The time it will take to do this depends on the state of charge of the batteries, and how much sun the array is getting.

General use and maintenance sequence

In general use, to turn the system on:

1. First throw toggle S2 to connect the battery bank to the output of the charger.
2. Then, throw toggle S1 to allow current to flow to the charger from the panel array.

If you do not follow this sequence you can destroy the charger.

When turning the unit off:

1. Throw toggle 1 to off position
2. Then throw toggle 2 to the off position.

Always follow these sequences for off/on operation. The toggles will effectively stop the flow of current to and from the chargers in the DC service box so that maintenance checks can be performed and components serviced. Always remember, however, that there is still power connected to the terminal blocks in the DC box at the input and output points.

Although high current toggle switches were used to cut off the positive input and output, a better choice would be high current DC rated 2 pole fused disconnect boxes for both panel input and DC box output.

Equalization charge

An equalization charge should be performed at least monthly during periods of diminished sun and at least every 90 days for a backup system such as Solar II. The frequency and duration of equalization charging will depend on how heavily the system is used. If it is used frequently and the batteries get drawn down quite a bit, equalize at least every 30 days; and for very heavy use, once a week might be appropriate. For normal operation, the duration of the equalization charge should be one to two hours.

When performing an equalization charge, always be present to observe and monitor the process, and pick a good sunny day.

Charge the batteries up first so that the voltmeter indicates about 14.5 volts before initiating an equalization charge.

If using CC1:

1. Throw the switch on the board to equalize to perform the equalization charge. The voltmeter should read about 15.5 volts.
2. When finished, throw the switch to the off position.

If using CC2:

1. Turn toggle one to the off position
2. Throw the knife switch to the 15 volt Zener/equalize position
3. Turn toggle one back on to the throw 2 position to energize CC2. The reading on the voltmeter should show about 15.5 volts.

For CC2, when the equalization charge is finished

4. Turn toggle one to the off position
5. Throw the knife switch to the 14 volt normal charging position
6. Turn toggle one back to the on position or throw 2.

After performing the equalization charge, check the electrolyte level of each cell in the battery bank and see if distilled water should be added. Do not overfill with water. During the evening, or after about six hours of no charging activity and no use of the batteries, check the specific gravity and ascertain whether another and/or longer equalization charge is needed the next day.

Further considerations

Depending on where you live, you may or may not have to conform to specific codes or regulations for setting up a photovoltaic system. You will have to seek the advice of the AHJ (authority having jurisdiction) for your country, state, province, town, to learn which components and set up procedures are allowed.

For the USA, NFPA 70 (also known as the NEC or National Electrical Code and specifically, article 690) contains most of the information needed for general code compliance information. However, the application of this code by the AHJ for your area could vary considerably so check locally for the details. Installations that satisfy code requirements will also satisfy mortgage companies, insurance companies, and local inspectors if this is an issue.

NFPA code sources and relevant material are listed in the URL section.

Appendix

Recommended reading

Solar Electricity by Simon Roberts (out of print)

The Solar Electric House by Steven J. Strong

Practical Electronics For Inventors by Paul Scherz

Any books by Forrest Mims – Excellent basic electricity and electronics books

Online resources

Lessons In Electric Circuits by Tony R. Kuphaldt http://openbookproject.net/electricCircuits/	Basic electricity and electronics
U.S. Solar Radiation Resource Maps http://rredc.nrel.gov/solar/old_data/nsrdb/1961-1990/redbook/atlas/	Insolation tables and maps with tilt corrections
Solar Radiation Resource Information http://www.nrel.gov/rredc/	PVWATTS calculator software
Solar Radiation and PV maps for EU countries http://sunbird.jrc.it/pvgis/countries/europe.htm	Insolation maps for EU countries
Homer http://www.nrel.gov/homer/	Calculator software
RET Screen International http://www.retscreen.net/ang/menu.php	Global insolation software
Database of incentives for renewable energy http://www.dsireusa.org	Information about various incentive programs to promote renewable energy and energy efficiency in the USA
Photovoltaic Systems Assistance Center http://www.sandia.gov/pv/	General resource for PV enthusiasts

Online resources, continued

SWRES http://www.nmsu.edu/~tdi/Photovoltaics/PV-Energy.html	General resource for PV enthusiasts
Home Power Magazine http://www.homepower.com/	General resource for PV enthusiasts
National Geophysical Data Center http://www.ngdc.noaa.gov/geomag/declination.shtml	Magnetic declination data
NFPA http://www.nfpa.org	NFPA 70 National Electrical Code

Resources for materials and components

McMaster Carr http://www.mcmaster.com	Many, many useful items
Surplus Sales Of Nebraska http://www.surplussales.com	Electronic and electrical supplies, heat sinks
Surplus Center http://www.surpluscenter.com	Toggle switches, and other electrical components, heat sinks
Circuit Specialists http://www.web-tronics.com	Electrical components
American Micro Semi Conductor http://www.americanmicrosemi.com	Electrical components
Mouser Electronics Inc. http://www.mouser.com/	Electrical components
Fair Radio Sales http://www.fairradio.com	Electrical components
Radio Shack http://www.radioshack.com	Electrical components
Backwoods Solar Electric Systems http://www.backwoodssolar.com	Solar electrical components

Kansas Wind Power http://www.kansaswindpower.net	Solar electrical components
New England Solar Electric Inc. http://www.newenglandsolar.com	Solar electrical components
AEE Solar http://www.aeesolar.com	Solar electrical components
The Solar Biz http://www.thesolarbiz.com	Solar electrical components
Independent Power & Light http://www.independentpowerllc.com	Solar electrical components and system installation
Trojan Battery Co. http://www.trojanbattery.com	Deep cycle batteries
Sargent-Welch http://www.sargentwelch.com	Knife switches
Filnor Inc. http://www.filnor.com	Knife switches
Cantex Inc. Http://www.cantexinc.com	Combiner and other electrical boxes
The RF Connection http://www.therfc.com	Red and black zip wire
Solorb / CirKits http://www.solorb.com/elect/solarcirc/scc3/index.html	12 volt, 20 amp charge controller kit
Delta Lightning Arrestors http://www.deltala.com	Lightning arrestors
Red Rock Energy http://www.redrok.com/main.htm	Tracker info

Other titles of interest

Build Your Own Solar Panel

by Phillip Hurley

Whether you're trying to get off the grid, or you just like to experiment, **Build Your Own Solar Panel** has all the information you need to build your own photovoltaic panel to generate electricity from the sun. The new revised and expanded edition has easy-to-follow directions, and over 150 detailed photos and illustrations. Materials and tools lists, and links to suppliers of PV cells are included. Every-day tools are all that you need to complete these projects.

Build Your Own Solar Panel will show you how to:

- Design and build PV panels
- Customize panel output
- Make tab and bus ribbon
- Solder cell connections
- Wire a photovoltaic panel
- Purchase solar cells
- Test and rate PV cells
- Repair damaged solar cells
- Work with broken cells
- Encapsulate solar cells

Available in print from Amazon.com and for download in full color PDF ebook format at

www.buildasolarpanel.com

*Download a free sample of **Build Your Own Solar Panel** in full color PDF at buildasolarpanel.com.*

Other titles of interest

E-books from Wheelock Mountain Publications:

Other titles by Phillip Hurley:

Build Your Own Solar Panel

Build a Solar Hydrogen Fuel Cell System

Practical Hydrogen Systems

Build Your Own Fuel Cells

The Battery Builder's Guide

Solar Supercapacitor Applications

and also:

Solar Hydrogen Chronicles edited by Walt Pyle

Tesla: the Lost Inventions by George Trinkaus

Tesla Coil by George Trinkaus

Radio Tesla by George Trinkaus

www.buildasolarpanel.com

Wheelock Mountain Publications
is an imprint of

Good Idea Creative Services
324 Minister Hill Road
Wheelock VT 05851

33118205R00110

Made in the USA
Lexington, KY
13 June 2014